Hope in the Anthropocene

# Hope in the Anthropocene

*Agency, Governance and Negation*

Edited by
Valerie Waldow, Pol Bargués and
David Chandler

EDINBURGH
University Press

Edinburgh University Press is one of the leading university presses in the UK. We publish academic books and journals in our selected subject areas across the humanities and social sciences, combining cutting-edge scholarship with high editorial and production values to produce academic works of lasting importance. For more information visit our website: edinburghuniversitypress.com

© editorial matter and organisation Valerie Waldow, Pol Bargués and David Chandler, 2024, 2026
© the chapters their several authors, 2024, 2026

Edinburgh University Press Ltd
13 Infirmary Street
Edinburgh EH1 1LT

First published in hardback by Edinburgh University Press 2024

Typeset in 10.5/13 Bembo by
IDSUK (DataConnection) Ltd, and
printed and bound by CPI Group (UK) Ltd,
Croydon, CR0 4YY

A CIP record for this book is available from the British Library

ISBN 978-1-3995-2985-3 (hardback)
ISBN 978-1-3995-2986-0 (paperback)
ISBN 978-1-3995-2987-7 (webready PDF)
ISBN 978-1-3995-2988-4 (epub)

The rights of Valerie Waldow, Pol Bargués and David Chandler to be identified as the Editors of this work have been asserted in accordance with the Copyright, Designs and Patents Act 1988, and the Copyright and Related Rights Regulations 2003 (SI No. 2498).

# Contents

List of Figures vii
Notes on Contributors viii

    Introduction: Dark Hope in the Anthropocene 1
    *Valerie Waldow, Pol Bargués and David Chandler*

## Part I Agency

1. The Anthropocene and the Unseen: Speculative, Pragmatic and Nihilist Hope 21
   *David Chandler*

2. A Hope Against Hope: Scandal, Cynicism and Critique in the Wake of the COVID-19 Polycrisis 37
   *Chris Zebrowski*

3. Working for 'Minor Utopias': Youth Employment in Sierra Leone and Liberia 51
   *Sukanya Podder and Raúl Zepeda Gil*

4. Visualising Hope in the Radical Data Work of W. E. B. Du Bois 68
   *Kiran K. Phull*

5. A Feminist Ethic of Care for Orienting Utopia in Adjuntas, Puerto Rico 86
   *Christie Nicoson*

## Part II Governance

6. Enduring Hopelessness: Governance without Horizon in Pandemic Times 103
   *Nicolas Gäckle*

7. Securing the Hopeful Subject? The Militarisation of Complexity
   Science and the Limits of Decolonial Critique — 119
   Claes Tängh Wrangel

8. The Hope–Colonialism Nexus — 135
   Marjo Lindroth and Heidi Sinevaara-Niskanen

9. Hopeful Times, Black Futures, and Things Quantum
   Technologies Tell about International Institutions — 150
   Geoff Gordon

10. In the Breaches of Cancelled Futures: The Entropies of
    Modernisation and Ecological Recomposition — 167
    Renan Porto

## Part III Negation

11. Hope and the End of Critique? Crisis and Affirmation in the
    Anthropocene — 185
    Valerie Waldow

12. Hope in a World That Will Never End? The Problem of
    Fanatical Hope in Critical Dystopias — 201
    Aristidis V. Agoglossakis Foley

13. Hope Makes Strange: Affect, Hope and Strangeness — 217
    Srishti Malaviya

14. Reimagining Hopeful Anthropocene Futures: From Entanglements
    to Radical Openness — 231
    Ignasi Torrent

15. Hope as a Theopolitical Virtue: Eschatology and End-of-Time
    Politics — 247
    Vassilios Paipais

Hope: An Epilogue — 262
Fleur Johns

Index — 266

# Figures

| | | |
|---|---|---|
| 4.1 | Photograph of the African American exhibit at the Paris Exposition, 1900 | 73 |
| 4.2 | 'City and Rural Population, 1890', W. E. B. Du Bois, 1900 | 75 |
| 4.3 | 'Proportion of Freemen and Slaves Among American Negroes', W. E. B. Du Bois, 1900 | 77 |
| 4.4 | *The Georgia Negro: A Social Study*, W. E. B. Du Bois, 1900 | 79 |
| 10.1 | Father. Photo by Renan Porto. | 177 |
| 10.2 | Tupinambá mantle. Created by Glicéria Tupinambá | 179 |

**Table**

| | | |
|---|---|---|
| 5.1 | A feminist ethic of care perspective on orientations of utopia | 94 |

# Notes on Contributors

**Pol Bargués** is Senior Research Fellow at CIDOB (Barcelona Centre for International Affairs) in Spain. Over the years he has developed an interest in the intersection of philosophy, critical theory and International Relations. In particular, he has critically interrogated international interventions in conflict-affected societies and explored the increasing prevalence of the ideas of resilience, hybridity and hope.

**David Chandler** is Professor of International Relations at the University of Westminster, UK. He edits the open access journal *Anthropocenes: Human, Inhuman, Posthuman*. His recent books include *The World as Abyss: The Caribbean and Critical Thought in the Anthropocene* (2023), *International Relations in the Anthropocene: New Agendas, New Agencies and New Approaches* (2021) and *Anthropocene Islands: Entangled Worlds* (2021).

**Aristidis V. Agoglossakis Foley** is a PhD candidate at the University of St Andrews, Scotland. His doctoral research adopts an interdisciplinary approach to explore the connections between dystopian literature and political theory, aiming to construct critical dystopianism. Drawing from political theory and dystopian thought, it is a critical and analytical tool of contemporary socio-political norms, practices and events.

**Nicolas Gäckle** is a PhD candidate in History and Theory of International Relations at the University of Groningen, the Netherlands. His research interests include the innovations and limits of the biopolitical as well as the experience of time in the Anthropocene and its governance. His work has previously been published in *Mobilities*.

**Geoff Gordon** is Senior Researcher in public international law at the Asser Institute, the Netherlands. In recent years, his research has centred on the impact of advanced infrastructural and information technologies on international institutions and practices. He has also focused on ecologies of computational technologies with machine learning or AI components, as well as the emerging field of quantum information technologies.

**Fleur Johns** is Professor in the Faculty of Law & Justice at UNSW Sydney, Australia, specialising in international law, legal theory, and law and technology. She is an Australian Research Council Future Fellow and a Visiting Professor at the University of Gothenburg, Sweden. She has authored five books, including her latest work titled *#Help: Digital Humanitarianism and the Remaking of International Order* (2023).

**Marjo Lindroth**, PhD, is Senior Researcher in the Arctic Centre at the University of Lapland, Finland. Her research critically investigates the intersections between rights, hope and contemporary colonialism.

**Srishti Malaviya** is Lecturer at the Centre for Writing Studies at O.P. Jindal Global University, Sonipat, India. Srishti holds a PhD in International Politics and her research interests include process philosophy, affect theory and posthuman and relational methodologies in international relations.

**Christie Nicoson** is a PhD candidate in Lund University's Department of Political Science and Agenda 2030 Graduate School, Sweden, where she studies gender, peace and climate change. Her current research departs from an intersectional feminist perspective to explore visions and enactments of peace in a changing climate.

**Vassilios Paipais** is Senior Lecturer in International Relations at the University of St Andrews, Scotland. He is the author of *Political Ontology and International Political Thought: Voiding a Pluralist World* (2017) and the editor of *Theology and World Politics: Metaphysics, Genealogies, Political Theologies* (2020) and *The Civil Condition in World Politics: Beyond Tragedy and Utopianism* (2022).

**Kiran K. Phull**, PhD, is Lecturer in International Relations based in the Department of War Studies at King's College London, UK. Her research focuses on global knowledge production, data politics, and the rise of scientific opinion polling, questioning how epistemic technologies create and shape the conditions for governing political subjecthood, and how innovative methods can help us to overcome the datafication of social life.

**Sukanya Podder**, PhD, is Reader in Post-War Reconstruction and Peacebuilding at King's College London. Her latest book is *Peacebuilding Legacy: Programming for Change and Young People's Attitudes to Peace* (2022).

**Renan Porto** is a writer and PhD researcher in law at the University of Westminster, UK, where he works on the emergence of spatial justice around the context of cacao production in northeast Brazil. He is author of the books O Cólera A Febre (2018) and Políticas de Riobaldo (2021).

**Heidi Sinevaara-Niskanen**, PhD, is Senior Lecturer and feminist scholar at the University of Lapland, Finland. Her critical research addresses the problematics of politics, hope and social sustainability.

**Ignasi Torrent** is Senior Lecturer in Politics and International Relations at University of Hertfordshire, UK, and a research member of the Critical Humanities and International Politics research group (CHIP). His research interests are framed around Critical Peace and Conflict Studies, the Anthropocene and new materialisms. He has published Entangled Peace (2021) and several articles in international relations journals.

**Valerie Waldow**, PhD, is Lecturer at the Department of Political Science at Otto-von-Guericke University of Magdeburg in Germany. Her research interests include international political theory, rationalities of international interventions and governance, Anthropocene discourses, and prospects for hope and critique in International Relations.

**Claes Tängh Wrangel** is Director of Centre for Multidisciplinary Studies on Racism (CEMFOR), Uppsala University, Sweden. His research centres on biopolitical security governance in and of liberal societies, focusing in particular on the use of hope as a biopolitical technology, as well as on the appropriation of neurobiology in US military discourses to govern complexity.

**Chris Zebrowski** is Senior Lecturer in Politics and International Relations and Director of the Centre for Security Studies (CSS) at Loughborough University, UK. His research investigates the historical evolution of the logics and techniques employed to govern crises. He is author of numerous peer-reviewed journal articles and The Value of Resilience: Securing Life in the Twenty-First Century (2016).

**Raúl Zepeda Gil** is a PhD candidate in the School of Security Studies at King's College London, researching youth involved in the Mexican drug war from a political economy perspective. Before his PhD Raúl studied political science at El Colegio de México and the National University in Mexico (UNAM). His research interests include criminal violence, conflict, peacekeeping, inequalities and climate change disasters.

# Introduction

## Dark Hope in the Anthropocene

*Valerie Waldow, Pol Bargués and David Chandler*

**Introduction**

Hope has come into its own in an unkind world, in which many experience a condition of loss and disorientation, while multilateral institutions and traditional forms of governance and cooperation appear unfit for purpose (Pupavac and Pupavac 2020). In times haunted by compounded crises, often expressed and brought together in differing ways (see Hine 2023; Zebrowski in Chapter 2), the politics of hope has come centre stage. Having hope today is not so much about an optimistic attitude or disposition, an assumption that everything will work out in the end (Snyder and Peterson 2000). To wish for the best in current circumstances would be akin to climate change denial or to denying entangled responsibilities to others (Chandler 2019). While some authors emphasise hope as an attitudinal resource enabling individual and collective human agency (for example, Brei 2016; Head 2016; McGeer 2004; McKinnon 2014; Moellendorf 2022; Oram 2016; Raygorodetsky 2017), hope has increasingly been reimagined and repurposed.

Most of these reimaginings seek to critique hope as optimism, yet they also salvage hope and put it to work in new forms (Bargués-Pedreny and Martin de Almagro 2020; Lindroth and Sinevaara-Niskanen 2022). This hope works against naive or desperate hopes in futurity as much as against discourses of hopelessness and resignation (Machado de Oliveira 2021, 241). For Rinaldo Walcott, hope, if it remains within a 'liberal progress narrative', would 'tie Black people to the regimen of slave and plantation logics' (2021, 2–3). Lauren Berlant (2011) argues that people are affectively attached to optimistic desires and ideals that are then not delivered by the current institutions, relationships and lifestyles, trapping people in cycles of disappointment and frustration.[1]

Hope, as traditionally framed, may be critiqued but often the critique provides for hope's recovery or repurposing in new or different forms. For Walcott and others, for example, this takes the form of the search and struggle for a 'new humanism' or a 'new genre of the human' (2021, 7; see also McKittrick 2015). For others, hope requires new actors and agencies such that collective hope is taken away from regulatory institutions or unrealistic desires. Sara Ahmed, for example, argues that hope is necessary as long as we maintain the element of uncertainty, deferral and promise in our imaginings, as long as we are open to possibility rather than closed to it (Ahmed 2010). 'Good hope', argues Aidan Hehir, 'necessitates determining whether the goals we hope to realise are actually achievable but also whether the strategy we plan to implement to achieve these goals is prudent' (2023, 217). If detached from neoliberal, colonial or capitalist imaginaries, hope emboldens agency and orients people towards a less certain set of generative possibilities (Head 2016).

Of increasing importance across the social sciences and humanities is a darker, more ontological, approach to hope – engaged with by several contributions to this book. This approach begins with the experience of a loss of world, an ontological loss, and insists that there is no straightforward claim to a future; that there is no easy exit or escape from our contemporary condition (Bliss 2015; Machado de Oliveira 2021). From catastrophic climate change to brutal episodes of mass migration, land dispossession or (post-)colonial anti-Black violence, this loss today is ontological, irretrievable. Everything that seemed to be good about modernity as scientific and technical progress, new forms of production and the rich extension of consumption choices, in the Anthropocene, is transvalued as negative and destructive. Histories of modernity's successes in the distant or even the recent past – of humanitarian development, democratising and state-building agendas – now appear in a new light as a litany of crimes of colonialism, extractivism, waste and hubris (Latour 2013, 76–7; see also Colebrook 2017, 16). It is recognised that these discourses of hope in relation to the Anthropocene, used as shorthand for the end of the world 'as we know it' in modernity, come from a place of privilege (Hine 2023, 34). The end of the world as we know it poses a crisis for those Vanessa Machado de Oliveira describes as engaged in 'low-intensity' struggles (2021, 52–3), those who have bought into and have gained from modernity at the expense of those who have been excluded. For those excluded and whose worlds have been forced to end to pay the price for the maintenance of modernity, the questions of ontological loss are not new.

If the question of hope in the Anthropocene is a pressing question, it is because it is a concern for the moderns, those for whom the Anthropocene or catastrophic climate change is the 'first crack' in the edifice that is the modernist, or modernist-colonial, ideology (Hine 2023, 17; Machado de Oliveira 2021). The moderns have increasingly come to the realisation that what has come to an end is merely the Western capacity to outsource 'the end of the world' to others and maintain their own myths of civilisation and progress (Hine 2023, 33; see also Colebrook, 2023). The Anthropocene is, then, the coming home of the consequences of the modernist disavowal of its rapacious devouring of life on earth, much as Aimé Césaire (2000, 36–7) and others saw fascism and the Holocaust as revealing the disavowed racial barbarism at the heart of coloniality. In this way, the Anthropocene opens up modernity to a critique that goes beyond any question of salvage or mitigation. The question that ontological loss poses, in this case, is how to live on in the realisation of our inescapable complicity in the disastrousness of this world (Shotwell 2016).

Once we recognise that the end of the world is a process of the ontological unravelling of modernity, new shoots of 'dark hope' appear, like the matsutake mushroom that grows in the ruins of capitalism (Tsing 2015). As Jairus Grove notes, 'the end of the world is not the end of everything' (2017, 205). Similarly, for the Dark Mountain Manifesto writers Paul Kingsnorth and Dougald Hine, 'The end of the world as we know it is not the end of the world full stop. Together, we will find the hope beyond hope, the paths which lead to the unknown world ahead of us' (Kingsnorth and Hine 2009). For Carl Cassegård and Håkan Thörn (2018, 563), this is the 'paradox of hope', which provides the wellspring for a postapocalyptic politics which arises in response to this loss of world. Decolonial theorist Vanessa Machado de Oliveira's well-cited book *Hospicing Modernity* (2021) captures this zeitgeist well, in constructing hope not in some desperate acts of salvation but in the opportunity for a generative disenchantment or disinvestment, as we first own and then move beyond our 'addiction' to the false promises of modernity.

Thus, ontological loss, with the supposition that we still must live on, places hope centre stage and opens up three fundamental areas of problematisation: those of agency, governance and negation. The first, that of agency, opens the problematic of how to orient ourselves and act in a world that cannot be rebuilt. The second area, governance, concerns questions of regulatory order and legitimation and the possibility that hope can be misappropriated in policy approaches of crisis management and disaster resilience. Finally, there is a third area of problematisation, at the centre of which is precisely the question of negation, where hope lies in the tasks of disruption, deconstruction and decomposition rather than in salvage and redemption.

This edited book brings together leading academics and junior scholars across the fields of politics, law, literary studies, geography and international relations to reflect on hope in the Anthropocene and the impact it has on our understandings of agency, governance and negation. The volume builds on several encounters among the authors, including a workshop convened by the editors at the European International Studies Association in Thessaloniki, Greece, in July 2022.[2] In these meetings, we shared our observations on the increasing centrality of hope and, particularly, on its darker notes after the end of modernity.

**Hope In Critical Theory**

In the traditional telling, where hope is cast as a relentless and naive optimism, hope has its origin in Greek mythology. The enigmatic lines of Hesiod, who explained how Zeus placed hope in Pandora's box next to all the other evils, have led contemporary scholars to debate whether hope was considered good (could it accompany humanity enabling a stand against its miseries?) or bad (as another evil that pushed man towards idleness and indolence). But in this introduction, we consider how to analyse hope in the Anthropocene through a heuristic focus upon the writings of theorists who struggled with the question of how to draw upon hope in a condition of fundamental, ontological, loss. For the critical theorists around the Frankfurt School, ontological loss arose from the immediate experience of the Holocaust as the fundamental revelation of the dark reality of the Enlightenment and its tenets of rationality, progress and universality.[3] For these theorists, the Holocaust was not just a terrible event that happened; rather, it implied the collapse of a whole framework of meaning and of a cultural framework of possibility; it had overthrown everything humanity could believe in (Adorno 1958).

The Holocaust represented a tipping point, in which modernity as a paradigm of 'enlightenment', 'progress' and 'development' was forced to be radically unthought. The means to engage in a critique of Enlightenment reason – culture, philosophy, science and politics – were also compromised (Burdman 2021). For Adorno, Auschwitz questioned the grounds of any meaningful socio-political or ethical engagement in the light of an irreversible past; precisely because the past cannot be undone, nothing, not even culture and critique, remained untouched by this (Adorno 2005, 107) – hence his famous dictum that 'to write poetry after Auschwitz is barbaric' (1983, 34), or Walter Benjamin's similar reminder that 'there is no document of civilisation which is not at the same time a document of barbarism' (Benjamin 1969, 256). What was the diagnosis of this barbaric civilisation?

Critical theorists of the mid-twentieth century pointed to the pursuit of knowledge and reason that had increasingly estranged humanity from nature. One of the core texts of the Frankfurt School, the *Dialectic of Enlightenment*, identifies the human quest to rule over nature and turn it into a pure object as the hubris that leads to catastrophe and downfall (Adorno and Horkheimer 1997). It was the 'levelling domination of abstraction' that enabled the Enlightenment project to produce the unreason which triggered industrial extermination (1997, 13). In a relentless polemic against human mastery, Adorno and Horkheimer cast light on the dialectical dynamics of the growing domination of nature and its effects on social relations. Humanity's drive to emancipate itself from natural forces did not lead to a state of reason and harmony; instead it led to subjection to techno-instrumental rationality, establishing a dysfunctional context of fear, self-alienation and the destructive domination of nature.

They critiqued a mode of thinking that takes itself to be superior to what it attempts to comprehend and disregards the fact that objects cannot be abstracted from social relations. This approach ultimately birthed two interrelated forms of violent abstractionism: positivism and scientific quantifying, on the one hand, and bureaucratic, disengaged ways of being and acting, on the other (Adorno 2005, 127; see also Lijster 2017). Already, for these early critical theorists, there was an intimate link between the modern scientific manner of interacting with nature and consequential physical destruction and loss, which reached a climax in the Holocaust, when Enlightenment turned against itself. 'Yet the fully enlightened earth radiates disaster triumphant', as the famous second sentence of the *Dialectic of Enlightenment* reads (Adorno and Horkheimer 1997, 3). This attack on modern scientific reason, as intensifying social alienation and fuelling the drives of catastrophic destruction, resembles the diagnosis made by Anthropocene thinkers today. While the world has ended, life nonetheless continues to go on in a broken tradition. Even after identifying that we are the problem, we are still here. The question arises: what do we do, then, in a broken world, faced with the fact that our lives (imagined as lived in accordance with civilisation and progress) were based upon self-deceit? What can we count upon when our ethical and political traditions are entirely complicit and compromised? The answer, then as now, is hope.

For critical theorists, hope is necessary in order to live on and enable any idea of meaningfully living a life (Jütten 2018, 285). It is necessary because it offers a way to think without being tainted by modernist constructions of rationalism and universalism; it nurtures a way of imagining the very possibility that the world could be different. In *The Principle of Hope* (Bloch 1995), widely read as the core statement on hope in the critical theory tradition (Browne 2005), Bloch describes hope as an attitude related to the 'not-yet-conscious'. His central point is directed against Sigmund Freud and his followers, who read the unconscious as a repressed past, thereby lacking a notion of the unconscious-as-hope, directed towards the future.

Hope is a basic trait of the human condition and is expressed and hidden in music, dreams and fairy tales. It links us to that which has not yet come into appearance. It is a capacity to dream forward, to daydream and feel that there is an objective truth, an objective world, of which we are not yet conscious but of which we might be conscious if we follow our intuition. Through intuition and sense, rather than reason, Bloch and other critical theorists pursued the possibility of reaching out to that which was objectively there but could not be grasped by our conscious reason and imagination (see further Jütten 2018).

Critical accounts differ, while sharing the understanding that hope neither depends on abstract wishful or positive thinking nor upon empirical affirmation of actually existing possibilities (Fuchs 2022; Gatens et al. 2021). While Bloch focused on the development potentialities of the future from the standpoint of the present, Adorno attempted to engage with the so far not yet lived potentialities of human life and history in the midst of a catastrophe through a radical critique of the present. After the Holocaust, it became implausible to hope for salvation by appealing to any fictional absolute. Consequently, the need arose to face up to the contemporary condition of irreparable failure. In doing so, Adorno articulates a distinct framing of hope, as a conception of liminality, existing only virtually, as potentiality, and in this way figuring a link to that which is radically other, beyond this barbaric world bereft of metaphysics (Adorno 2005, 229). Because it does not exist within the world of modernity, hope is a force of negation, operating outside and otherwise to the very structures and thinking that made the Holocaust possible (see a similar figurative perspective of 'negativation' put forward in Ferreira da Silva 2022). Any positive reference point for hope within this world would necessarily entail a relapse into totalitarian ideology because it must draw on our compromised sources of thought and imagination. However, to dismiss hope would be equally barbaric. In the end times of the modernist world, hope remains the only possible strategy for a negative approximation to a different, just or true reality, to a different possible future.

Like Bloch, Adorno did not seek hope in the sublime or in otherworldly experiences; as Chua (2004, 525) explains, 'the "non-being of hope" must nevertheless be a this-worldly metaphysics, an otherness that is material and transient, a real experience and not some supersensible or supernatural idea'. For Adorno, hope is found in the particular, in the 'transfigured experience' of children playing without purpose and abstraction:

> The little trucks travel nowhere . . . The tiny barrels on them are empty . . . However mistaken the child may be, he is able to perceive the *nonidentical* in

what will become a grown-up world where these names will be subsumed under abstract concepts and reified into commodities. (Chua 2004, 526–7)

For Adorno, then, hope cannot lie 'in' the world ontologically, but nevertheless exists in the potentiality of the present, that is, beyond the grasp of conceptual representation. We feel this is summarised best in his statement, towards the end of his magisterial work *Negative Dialectics*:

> To Beckett, as to the Gnostics, the created world is radically evil, and its negation is the chance of another world that is not yet. As long as the world is as it is, all pictures of reconciliation, peace, and quiet resemble the picture of death. The slightest difference between nothingness and coming to rest would be the haven of hope, the no man's land between the border posts of being and nothingness. (Adorno 2007, 381)

What unites these early critical theorists, therefore, is an awareness that is very close to that of the theorists of the 'dark hope' of the Anthropocene; an awareness that hope cannot lie in operations of salvation or of rescue or of returning to some past 'status quo' of imagined 'sustainability'. The world as we knew it is, or needs to be, over; and, as Machado de Oliveira notes, the only task is to wean ourselves off our dependencies and addiction to that world's false promises of autonomy, of human exceptionalism, and of endless progress (2021). There is nothing that can be salvaged from modernist ontologies of cuts and separations and human exceptionalism. From this perspective, the whole tradition of modernist political thinking is tarnished and to make any contemporary politics legible within it would entail reproducing modernity's violence and unsustainability (Machado de Oliveira 2021, 181). This ethico-political register of fundamental loss and of a reconfigured mode of hope underlies the diagnosis of and response to the Holocaust in ways that can inform analyses of hope in the Anthropocene.

## Dark Hope in the Anthropocene

In the Anthropocene, the realisation that humans have fundamentally impacted the Earth's geology and ecosystems has clarified that the problem is 'us', our ways of worlding: 'In worshipping the god of progress, we have unleashed the dogs of war, and it seems that the war dogs are us' (Bird Rose 2011, 10). The logic that was allegedly developing behind our backs for centuries now 'manifests itself in innumerable possible hairline cracks in one's familiar world and its weathers' (Clark 2013, 20). 'Agrilogistics', the strict separation of nature and culture in food production, writes Morton, is the 12,000-year-old agricultural programme that 'promises to eliminate fear, anxiety, and contradiction – social, physical, and ontological – by establishing thin rigid boundaries between human and nonhuman worlds and by reducing existence to sheer quantity' (Morton 2016, 43). In searching for certainty and security, agrilogistics produces new uncertainties and insecurities. 'Human being disturbs Earth and its lifeforms in its desperate and disturbing attempt to rid itself of disturbance' (Morton 2016, 64). That is, agrilogistics is compelling to humans and yet it is the source of all other 'wicked problems' – irreducible and thus complex to solve – such as industrialisation, accelerated agriculture or global warming: 'Agrilogistics is the smoking gun behind the smoking chimneys responsible for the Sixth Mass Extinction Event' (Morton 2016, 43).

Thus, according to narratives reflecting upon the Anthropocene, the end of the world is not a trope; it is real and has already occurred. Unlike other more traditional environmental perspectives, which demand action to prevent future disaster, the key break in the Anthropocene is the realisation that times are 'postapocalyptic', as the catastrophe is ongoing and is inevitable (Cassegård and Thörn 2018). 'The current extinction crisis is an Earth-shattering disaster, one that cannot be unmade', wrote Bird Rose (2011, 5). The Anthropocene is considered an 'event within knowledge and human history [that] alter[s] the relation between thought and its outside' (Colebrook 2017, 10). Timothy Clark argues that the Earth refuses to be a frame of meaning or a shared ground, 'the supremely taken-for-granted'; now even the rain turns into 'an event', 'an absolutely different singularity': 'the Anthropocene could be said to be marked by the fact that the earth itself, its weathers and its shared finite horizons of land, sea and sky, becomes newly astonishing in intellectually challenging and sometimes frightening ways' (Clark 2013, 6–7).

The geological forces of the Earth put into question the nature/culture divide and the binary logics that follow from it, thus breaking fundamentally with modernity, which ultimately assumed that the world could be read by humans (Hamilton et al. 2015). In this sense, the planetary challenge of the Anthropocene is considered to exhaust traditional approaches to international relations (Burke et al. 2016), conceptions of ecology and security (Fagan 2016) or mobility (Baldwin et al. 2019) that placed the Earth in the background, as opposed to recognising it as a collective assemblage of interactive agencies (Latour 2013, 2018). A central question is therefore posed: how do we hope in the ruins of modernity? In seeking to answer this question, some theorists of the Anthropocene look at how Indigenous peoples cope with the postapocalyptic present of colonial brutality and world devastation (Chandler and Reid 2019; Lempert 2018). Jonathan Lear (2008, 34) finds hope in the way the

Crow Nation survived the end of their way of life, when they were displaced to a reservation and their world had been taken away cosmologically. 'Radical hope', he writes, is a 'dream-vision' that gives courage to go forward even though 'we as yet lack the concepts with which to understand what we are reaching out for' (Lear 2008, 122).

A key realisation in the Anthropocene is that humans were never separated from nature completely. Neither colonial conquests and wars, nor scientific reason, industrial innovation and mass production, which attempted systematic distinctions and cuts, could finish with the world of entanglements. Indeed, the Anthropocene reveals that the opposite appears to be the case: 'We find ourselves like prisoners waking up inside the ecological mesh of lifeforms' (Morton 2013, 192). In these accounts, hope is not a desire for understanding, progress or salvation. Hope is akin to connectivity: hope that we do not become abandoned and alone in our sorrows, alienated from other beings; hope that our connections enrich and multiply (Haraway 2003). Thus, from the Holocaust to the Anthropocene, the end of the world entails hope. We are indeed 'here' in the world, enmeshed, connected, and the key gesture is to learn to care for the imbroglios, the 'monsters' that moderns have created (Latour 2011). Rather than killing other creatures and disposing of them as waste, 'it is a willingness toward dialogue, a willingness toward responsibility, a choice for encounter and response, a turning toward rather than a turning away' (Bird Rose 2011, 5).

A theorist who perhaps most exemplifies the 'dark hope' of the Anthropocene is Alexis Pauline Gumbs, particularly in her well-cited book *Undrowned: Black Feminist Lessons from Marine Animals* (2020). Here, there is no mourning for the loss of modernist imaginaries that cut us from relationships and our embedded ecologies. The end of the world is not merely a matter of struggle and resistance but a matter of shared care and love; a matter of learning to 'listen across species, across extinction, across harm' (2020, 15). Her book is one of hope as an ethical or poetical practice; a hopeful practice of unmaking or undoing the human, enabling work on the self that opens us up to ways of living and identifying with others. This is a book for 'everyone who knows that a world where queer Black feminine folks are living their most abundant, expressed, and loving lives is a world where everybody is free' (2020, 13). Freedom is not a matter of fighting the world, of mastering it, but of surrendering to it. Freedom is a matter of hope:

> AND WHAT HAPPENS IF WE just let go? Like dolphins who beach themselves on shore to eat, and trust the tide to bring them back into the water, or who time their birth cycles to seasonal floods, or migrate across the world following warm currents on a menopausal planet. What it would

take to tune in with our environment enough to be in flow with the Earth, instead of in struggle against it (2020, 121).

Hope has become transformed in the shift from modernist political and governmental imaginaries to more relational, interdependent and entangled understandings of agency, governance and negation. As the role of contingency and unintended and unforeseen consequences comes to the fore, narrow, linear understandings of universal rationality appear to have less purchase. This creates space for more flexible, more provisional and more imaginative forms of politics and governance, including gestures of refusal and negation, gestures which, as the Gesturing Towards Decolonial Futures collective emphasises, are not expected to necessarily save or transform the world but hope to fail generatively, 'preparing for the end of the world as we know it, and showing up differently so that "another end of the world" becomes possible' (GTDF n.d.). It is here that hope undergoes a radical or 'dark' transformation, from the margins to the centre of political thought. It is this transformation that is one of the central problematics of this book.

**Conclusion: A Guide to Dark Hope**

In this introductory chapter we have highlighted the significance of concerns about fundamental loss that are present in much of twentieth-century critical theory, in the wake of the Holocaust, to provide an analytical perspective for discussions of contemporary framings of hope. Analogies can be carefully drawn between the ways that critical theorists and many contemporary Anthropocene thinkers share a specific ontological problematic (namely, how to go on living after the discovery that the world has ended) and a critique of the imbrication of modernist reason within the machinic logics of extraction and extinction (see further Bargués et al. 2024). In so doing, they ground hope ontologically (as a liminal and immanent potentiality based on relations of becoming with unknowable others), foregrounding the problematics of agency, governance and negation. This edited collection focuses upon the shifting understanding of the ethics and politics and temporalities of hope in the Anthropocene and seeks to analytically engage a range of diverse writers in a discussion of how hope works differently today. The book is structured into three parts, with a concluding epilogue from Fleur Johns.

*Agency*

In the first part of the book, the chapters discuss how hope becomes a force that brings together and enables human agency. However, the authors do not associate hope with wishful thinking, eco-modernism, governmental promises or quick-fix solutions but with the bottom-up, emphatic practices and cultivation of new sensibilities rooted in everyday experiences. In Chapter 1, David Chandler discusses how hope now plays a pivotal role, emerging as a 'discursive field' seeking in different ways to access an 'unseen' realm beyond the grasp of liberal or Enlightenment 'reason'. Hope thus reveals the growing importance of ontology and the 'unseen' in social and political thought during the Anthropocene era. Chandler identifies three forms of agential practice within these discourses: speculative hope, relying on speculative imagination to act in the present with generosity and care; pragmatic hope, focusing on grounded experimentation and worldmaking practices; and nihilist hope, confronting modernity from its constitutive 'beyond' through negation.

In Chapter 2, Chris Zebrowski reflects on the dominant responses to the entangled multiplicity of polycrises in the Anthropocene. Zebrowski critiques modernist narratives that hope for a return to normal life as these have only led to apathy and cynicism, in turn cementing the crises of the present. Instead of a promise of rapid salvation, he encourages the retention of hope in a hopeless situation, while re-narrating and developing new tales that could nurture other and better ways of life. In Chapter 3, Sukanya Podder and Raúl Zepeda Gil explore hope as a participatory exercise in active imagining that young people in Sierra Leone and Liberia show in a context of economic difficulties and structural constraints. They contrast this hope with the hope of coping with daily hardships that is offered by top-down and adult-centred policies funded by international state-building programmes that employ youth in precarious jobs. Instead, Podder and Gil value enlarged approaches to hope, which promise a more long-term and futures-oriented programming that is rooted in intergenerational responsibility and the participation and inclusion of young people.

Challenging the risks associated with the datafication of modern life, Kiran Phull in Chapter 4 shows how data can also be used to craft a better world. She draws on W. E. B. Du Bois's visual statistical work, exhibited at the 1900 Exposition Universelle in Paris, which represented the global colour line and connected local anti-racist practices with global anti-colonial struggles. Du Bois's visualisations are bleak, ambiguous and expectant at the same time; they resist teleological narratives of progress, embody Black subjecthood, and map the Atlantic slave trade to produce, according to Phull, 'a hopeful act of anticolonial worldmaking'. In Chapter 5, Christie Nicoson relates the climate risks in today's Puerto Rico and other Caribbean countries with the past and present experience of colonialism. In a highly climate-vulnerable country with adverse social-political conditions and a neocolonial economy based on extractivism and tourism, Nicoson finds hope in the community group Casa Pueblo in Adjuntas, which resists open-pit mining projects and opts for self-governance and the building of alternative energy production projects – a concrete utopia, as she calls it, that pursues and unites an anti-colonial struggle, environmental justice and community-led conviviality.

*Governance*

The second part of the book explores how hope has increasingly entered and shaped policy agendas, sometimes being enrolled in governance programmes that curtail change rather than enabling it. Here a more ambivalent picture of hope is presented. Hope increasingly becomes an ambiguous or paradoxical concept, operating liminally: at times entirely captured as resource for governance and control and, on other occasions, constructed in alternative ways as a resource of survivance, resistance or refusal. In the first chapter of this section, Chapter 6, Nicolas Gäckle examines hope under the pandemic conditions of the governance of COVID-19. Here the question of the emergency posed itself less in the register of a hope for a quick return to normality and instead pointed to a need to endure a continuously extended present. This had the effect of undermining imaginaries of governance that had built their appeal on a progressively unfolding horizon towards which subjects could aspire. The experience of this temporal suspension, Gäckle argues, puts the availability of hope itself into question.

In Chapter 7, Claes Tängh Wrangel engages with discourses which locate hope onto-politically in understandings of complexity and relational entanglement, held to always be in excess of modernist mechanisms of representation and human/nature binaries. Wrangel argues that complexity sciences have been appropriated by the US military as a means to unlock the 'driving forces behind creation itself', reconfiguring the colonial matrix around a complex and neurobiological subject. Thus, hope becomes inculcated within new colonial forms of governance and control. This theme is also central to Chapter 8 where Marjo Lindroth and Heidi Sinevaara-Niskanen highlight how hope is instrumentalised within colonial mechanisms of control and the ways in which Indigenous survivance is repurposed towards the reproduction of colonial hierarchies of power. They also point to a paradox, that while the hope of Indigenous peoples is used to defer and suspend Indigenous demands and concerns, Indigenous modes of being are repurposed and imagined as capable of providing hope for settler-colonialism's survival in a world of global warming and climate catastrophe.

In Chapter 9, Geoff Gordon reads two framings of hope entangled via a quantum lens, articulated in the terminology of 'cruel optimism' and of 'hope draped in black'. The 'cruel optimism' of hope as articulated by the European

Commission's Quantum Flagship (EQF) initiative promises disruptive contingency but is constrained by its purpose of promoting continuity with established patterns of political-economic distribution. This is counterposed to the 'hope draped in black' of the Black Quantum Futurism (BQF) project, articulated via the barred Black subject, defiant, mired in life, and melancholic in a way that recalls the vitality of mourning songs and elegies in a Black tradition. In the closing chapter in this section, Chapter 10, Renan Porto draws on Indigenous cosmologies informed by his family background where cocoa is still planted using the traditional cabruca system of agroforestry around the city of Jequié in the state of Bahia, northeast Brazil. This – clearly counterposed to placing hope in technocratic governance discourses of direction and control, held to be able to sustain racial capitalism in the age of the Anthropocene – is a different form of hope, one based upon being present in more-than-human relations of ecological recomposition.

*Negation*

The third part of our book explores how hope can be understood as a more negative disposition towards the present. Here hope emerges through a struggle to fully recognise the extent of existential loss occasioned by ongoing histories of dispossession, ecocide, genocide, coloniality and racial capitalism. It is not possible to be in this world without being, to some extent, complicit in the situation of destruction and degradation. Thus, to possess any type of forward-looking hope would be a mode of disavowal, perpetuating the contemporary condition. In this context, hope must necessarily be an aspiration to undo what is, including self-understandings of the modernist subject; a hope to end this world so that a new beginning can come. In the opening chapter of this section, Chapter 11, Valerie Waldow explores the implications of this shift for critical analysis, suggesting that the revival of hope and the dismissal of critique in debates surrounding the Anthropocene are intertwined phenomena. Waldow examines how hope and critique are being redefined, replacing traditional utopian hope with 'radical hope' embracing new possibilities amid increased ontological vulnerability. However, this process has led to a closure of critical spaces. Waldow argues that hope remains necessary for critical intervention but needs to transcend wishful thinking or the affirmation of existing possibilities.

In Chapter 12, Aristidis Victor Agoglossakis Foley reads classic dystopian novels to reveal that they usually retain a lifeline of hope and salvation. Nevertheless, when promising hope amid the most depressive and despairing futures, Foley argues, dystopias' powerful criticism of society becomes undermined. We may end up as hopeful fools, Foley adds, 'hoping foolishly, blindly in the face of total destruction'. The risk is that hope becomes an uncritical, innocuous, desire; a passive expectation rather than being agentic and productive. Chapter 13 by Srishti Malaviya critiques the Anthropocene's tendency to emphasise finitude and crisis while neglecting continuity. At the core of her argument lies the exploration of the 'hope–strangeness relation' in its ongoing resistance to finite conceptions of the world. Malaviya draws upon affect theory, particularly Ben Anderson's framing of 'affective hope' (2014), to explore the 'uncanny strangeness' and indeterminate nature of the forces that shape becoming. By 'unmooring hope' from human subjectivity, Malaviya reveals that indeterminacy is not a new crisis arising from a human-centred world but an integral aspect of how experiences unfold.

In Chapter 14, Ignasi Torrent challenges an excessive focus on interconnectedness and presents a different perspective on hope in the Anthropocene, embracing radical openness and questioning deterministic assumptions of entangled ontologies. The chapter explores instances of non-relational or beyond-the-relational being that resist engagement with the entangled world, suggesting that non-engagement and withdrawal can be forms of political resistance and survival. Hope is seen as the possibility for unlimited arrangements driven by unpredictable forces enabling generative collisions and incompossibilities, hence exceeding established notions of generative possibility. In Chapter 15, Vassilios Paipais discusses hope as a theopolitical virtue in a nihilist era. Delving into the ideas of Jewish messianic nihilism and Christian Trinitarianism, he argues against separating eschatology from politics and exploiting it for political or technoscientific utopias. Instead, Paipais suggests that eschatological hope can transform politics into 'a counter-politics of happiness, resistance, messianic profanation, and theocratic an-archy'. Blurring the boundaries between theology and nihilism allows for critique and a reimagining of politics and human existence in an uncertain world shaped by apocalyptic urgency and technoscientific millenarianism.

In the epilogue, Fleur Johns reviews the chapters of the book to find two common threads. One is that hope begins with the assumption of loss and the acknowledgement that proceeding according to business as usual is not an option, since traditional, modernist methodologies have become obsolete in the Anthropocene. Another thread is how dark hope is oriented towards the present rather than towards the future, a present haunted by insecurities in which hope can be found in people's various practices and struggles, as much as in the negation of the stable, the normal, and the promises of betterment. Unlike usual epilogical scripts which are future-oriented and laden with promises of hopefulness, therefore, Fleur invites us to think of hope without a happy ending. As she wonders: in the Anthropocene, what might it entail to survive 'the wreck' without trying to secure the future?

## Notes

1. Similar concerns can be found in critical fantasy studies, drawing from the psychoanalytic theories of Klein and Lacan, stressing both how desires are governed through fantasies and also the worldmaking aspects of fantasies (Fletcher, 2019; Behagel and Mert, 2021).
2. Prior to this, the editors shared conversations on hope, the Anthropocene and critical theory, initially with Lucia Najšlová at the Institute of Political Studies, Charles University, Prague, in November 2018, and then with Sebastian Schindler at the 8th EISA Exploratory Symposia in Rapallo in October 2019. We would like to thank them both for their important insights and support for this project.
3. By highlighting these parallels we do not intend to equate the Anthropocene and the Holocaust or ignore other narratives of trauma and suffering by emphasising the singularity of either of these events. Instead, our focus is on the structurally similar problematisation of fundamental loss.

## References

Adorno, T. 1958. Einführung in die Dialektik. In C. Ziermann (ed.), *Nachgelassene Schriften*, Abteilung IV: *Vorlesungen*. Band 2. Frankfurt am Main: Suhrkamp.
Adorno, T. 1983. Cultural criticism and society. In *Prisms*. Cambridge, MA: MIT Press, 17–34.
Adorno, T. 2005. *Minima Moralia: Reflections from Damaged Life*. London and New York: Verso.
Adorno, T. W. 2007. *Negative Dialectics*. New York: Continuum.
Adorno, T. and Horkheimer, M. 1997. *Dialectic of Enlightenment*. New York: Verso Classics.
Ahmed, S. 2010. *The Promise of Happiness*. London: Duke University Press.
Anderson, B. 2014. *Encountering Affect: Capacities, Apparatuses, Conditions*. London: Routledge.
Baldwin A., Fröhlich, C. and Rothe, D. 2019. From climate migration to Anthropocene mobilities: Shifting the debate. *Mobilities* 14(3): 289–97.
Bargués-Pedreny, P., and Martin de Almagro, M. 2020. Prevention from afar: Gendering resilience and sustaining hope in post-UNMIL Liberia. *Journal of Intervention and Statebuilding* 14(3): 327–48.
Bargués, P., Chandler, D., Schindler, S. and Waldow, V. 2024. Hope after 'the end of the world': Rethinking critique in the Anthropocene. *Contemporary Political Theory*.
Behagel, J. H. and Mert, A. 2021. The political nature of fantasy and political fantasies of nature. *Journal of Language and Politics* 20(1): 79–94.

Benjamin, W. 1969. *Illuminations: Essays and Reflections*. London: Vintage.
Berlant, L. 2011. *Cruel Optimism*. Durham, NC: Duke University Press.
Bird Rose, D. 2011. *Wild Dog Dreaming: Love and Extinction*. Charlottesville: University of Virginia Press.
Bliss, J. 2015. Hope against hope: Queer negativity, Black feminist theorizing, and reproduction without futurity. *Mosaic: An Interdisciplinary Critical Journal* 48(1): 83–98.
Bloch, E. 1995. *The Principle of Hope*, vol. 1. Cambridge, MA: MIT Press.
Brei, A. T. 2016. *Ecology, Ethics and Hope*. London and New York: Rowman & Littlefield.
Browne, C. 2005. Hope, critique, and utopia. *Critical Horizons* 6(1): 63–86.
Burdman, J. 2021. 'After Auschwitz': Writing history after injustice in Adorno and Lyotard. *Contemporary Political Theory* 20: 815–35.
Burke, A., Fishel, S., Mitchell, A. et al. 2016. Planet politics: A manifesto from the end of IR. *Millennium: Journal of International Studies* 44(3): 499–523.
Cassegård, C. and Thörn, H. 2018. Toward a postapocalyptic environmentalism? Responses to loss and visions of the future in climate activism. *Environment and Planning E: Nature and Space* 1(4): 561–78.
Césaire, A. 2000. *Discourse on Colonialism*. New York: Monthly Review Press.
Chandler, D. 2019. The death of hope? Affirmation in the Anthropocene. *Globalizations* 16(5): 695–706
Chandler, D. and Reid, J. 2019. *Becoming Indigenous: Governing Imaginaries in the Anthropocene*. London: Rowman & Littlefield.
Chua, D. K. L. 2004. Adorno's metaphysics of mourning: Beethoven's farewell to Adorno. *The Musical Quarterly* 87(3): 523–45.
Clark, T. 2013. What on world is the Earth?: The Anthropocene and fictions of the world. *Oxford Literary Review* 35(1): 5–24.
Colebrook, C. 2017. We have always been post-Anthropocene. In R. Grusin (ed.), *Anthropocene Feminism*. Minneapolis: University of Minnesota Press, 1–20.
Colebrook, C. 2023. *Who Would You Kill to Save the World?* Lincoln: University of Nebraska Press.
Fagan, M. 2016. Security in the Anthropocene: Environment, ecology, escape. *European Journal of International Relations* 23(2): 292–314.
Ferreira da Silva, D. 2022. *Unpayable Debt*. London: Sternberg Press.
Fletcher, R. 2018. Beyond the end of the world: Breaking attachment to a dying planet. In I. Kapoor (ed.), *Psychoanalysis and the Global*. Lincoln: University of Nebraska Press, 48–69.
Fuchs, T. 2022. The not-yet-conscious: Protentional consciousness and the emergence of the new. *Phenomenology and the Cognitive Sciences*, https://doi.org/10.1007/s11097-022-09869-9.

Gatens, M., Steinberg, J., Armstrong, A., James, S. and Saar, M. 2021. Spinoza: Thoughts on hope in our political present. *Contemporary Political Theory* 20(1): 200–31.
Grove, J. 2017. The geopolitics of extinction: From the Anthropocene to the Eurocene. In D. R. McCarthy (ed.), *Technology and World Politics: An Introduction*. Abingdon: Routledge, 204–23.
GTDF (Gesturing Towards Decolonial Futures). n.d. Reciprocity commitments. *DecolonialFutures.net*. Available at https://decolonialfutures.net (accessed 21 December 2023).
Gumbs, A. P. 2020. *Undrowned: Black Feminist Lessons from Marine Animals*. Chico, CA: AK Press.
Hamilton, C., Bonneuil, C. and Gemenne, F. (eds). 2015. *The Anthropocene and the Global Environmental Crisis: Rethinking Modernity in a New Epoch*. Abingdon: Routledge.
Haraway, D. 2003. *The Companion Species Manifesto: Dogs, People, and Significant Otherness*. Chicago, IL: Prickly Paradigm Press.
Head, L. 2016. *Hope and Grief in the Anthropocene: Re-conceptualising Human–Nature Relations*. London: Routledge.
Hehir, A. 2023. 'An expensive commodity'? The impact of hope on US foreign policy during the 'unipolar moment'. *European Journal of International Relations* 29(1): 202–26.
Hine, D. 2023. *At Work in the Ruins: Finding our Place in the Time of Science, Climate Change, Pandemics & All the Other Emergencies*. London: Chelsea Green Publishing.
Jütten, T. 2018. Adorno on hope. *Philosophy & Social Criticism* 45(3): 284–306.
Kingsnorth, P. and Hine, D. 2009. *Uncivilisation: The Dark Mountain Manifesto*. Oxford: The Dark Mountain Project. Available at https://dark-mountain.net/about/manifesto/ (accessed 21 December 2023).
Latour, B. 2011. Love your monsters. *Breakthrough Journal* 2(11): 21–8.
Latour, B. 2013. *Facing Gaia, Six Lectures on the Political Theology of Nature: Being the Gifford Lectures on Natural Religion, 18th–28th of February 2013*. Edinburgh: draft version.
Latour, B. 2018. *Down to Earth: Politics in the New Climatic Regime*. Cambridge: Polity Press.
Lear, J. 2008. *Radical Hope: Ethics in the Face of Cultural Devastation*. Cambridge, MA: Harvard University Press.
Lempert, W. 2018. Generative hope in the postapocalyptic present. *Cultural Anthropology* 33(2): 202–12.
Lijster, T. 2017. 'All reification is a forgetting': Benjamin, Adorno, and the dialectic of reification. In S. Gandesha and J. F. Hartle (eds), *The Spell of*

*Capital: Reification and Spectacle*. Amsterdam: Amsterdam University Press, 55–66, www.jstor.org/stable/j.ctt1pk3jqt.6.

Lindroth, M. and Sinevaara-Niskanen, H. 2022. *The Colonial Politics of Hope. Critical Junctures of Indigenous–State Relations*. Abingdon: Routledge.

Machado de Oliveira, V. 2021. *Hospicing Modernity: Facing Humanity's Wrongs and the Implications for Social Activism*. Berkeley, CA: North Atlantic Books.

McGeer, V. 2004. The art of good hope. *Annals of the American Academy of Political and Social Science* 592: 100–27.

McKinnon, C. 2014. Climate change: Against despair. *Ethics & the Environment* 19: 31–48.

McKittrick, K. (ed.) 2015. *Sylvia Wynter: On Being Human as Praxis*. Durham, NC: Duke University Press.

Moellendorf, D. 2022. *Mobilizing Hope: Climate Change and Global Poverty in the 21st Century*. Oxford: Oxford University Press.

Morton, T. 2013. *Hyperobjects: Philosophy and Ecology after the End of the World*. London and Minneapolis: University of Minnesota Press.

Morton, T. 2016. *Dark Ecology: For a Logic of Future Coexistence*. New York: Columbia University Press.

Oram, R. 2016. *Three Cities: Seeking Hope in the Anthropocene*. Wellington, New Zealand: Bridget William Books.

Pupavac, V. and Pupavac, M. 2020. *Changing European Visions of Disaster and Development: Rekindling Faust's Humanism*. London: Rowman & Littlefield.

Raygorodetsky, G. 2017. *The Archipelago of Hope: Wisdom and Resilience from the Edge of Climate Change*. New York: Pegasus.

Shotwell, A. 2016. *Against Purity: Living Ethically in Compromised Times*. Minneapolis: University of Minnesota Press.

Snyder, C. R. and Peterson, C. 2000. *Handbook of hope: Theory, Measures and Applications*. San Diego, CA: Academic Press.

Tsing, A. L. 2015. *The Mushroom at the End of the World: On the Possibility of Life in Capitalist Ruins*. Princeton, NJ: Princeton University Press.

Walcott, R. 2021. *The Long Emancipation: Moving Toward Black Freedom*. Durham, NC: Duke University Press.

# Part I
# Agency

# 1

# The Anthropocene and the Unseen: Speculative, Pragmatic and Nihilist Hope

*David Chandler*

**Introduction**

In the Anthropocene, it appears that we live in a bifurcated world, with a small world of the 'moderns' and a much larger unseen and unacknowledged world beyond this. For example, the late Bruno Latour argued that the modern ontology was too restrictive, creating an artificial world separated from the world that exists in reality. The division was described by him as that between 'the world I live in as a citizen of a developed country, and . . . the world I live off, as a consumer of the same country' (Latour 2021, 41). For contemporary critical social theorists, the modern ontology (of the human as self-determined subject separate from a world composed of other-determined objects) is a problematic abstraction, failing to capture the complexity of real life. This failure means that we do not take into full account the exploitation of the natural environment. For Latour, it is 'as if every wealthy state was coupled with a shadow state that never stopped haunting it, a sort of Doppelgänger that provides for it, on the one hand, but is devoured by it, on the other' (Latour 2021, 41). A new way of living and interacting with nature is at stake in the Anthropocene and hope is central to it. But, of course, this is not the 'modernist' hope. Thus, hope (in the framing of this chapter) is not defined merely in terms of possessing a positive approach to future outcomes. Hope is not a subjective attribute or positive mental state but a discursive field of practices or activities designed to access what is present but is unseen. Hope is grounded upon a reality that exists not on the transparent surface of appearances but in unseen potentiality and thus beyond the world of liberal or Enlightenment 'reason'. Hope is connected thereby to the practice of living after the 'end of the world': after the exhaustion of the world as constituted in a modern ontology. This chapter will heuristically sketch out three distinct approaches to hope in the Anthropocene. However, first there will be a brief engagement with hope as it was presented in modernity.

**Hope in Modernity**

In modernity, hope was a marginal preoccupation. Hope operated outside the bounds of the world of politics and instrumental reason and was articulated in terms of the immanent power of life beyond the human realm of reason. It should be noted that these immanent powers were not generally seen as accessible to the modern subject. So, while two worlds existed – the world of modernity (of universal reason, the ontic, the actual, the world of Newtonian determinism and natural laws) and the world of immanent vital forces – there was understood to be a clear divide between them. This divide was articulated as that between the world of politics and law, of universal reason, and the world beyond, of natural and social and economic forces, that worked unseen behind the backs of reason or intention. Two of the most famous examples of this framing, in terms of being classics of modernist thought, would perhaps be Adam Smith's 'invisible hand' of the market (Smith 2022 [1776]) and Immanuel Kant's view of nature's 'secret plan' or 'providence' (Kant 1991).

This other world was veiled or curtained off from reason's access although it was understood to be a world that was the precondition for human betterment and improvement. Thus, for Kant, 'if we assume a plan of nature, we have grounds for greater hopes' (1991, 52) enabling thereby ethical and future-oriented actions regardless of the empirical experiences of disaster and defeat (see also Connolly 2011, 148). This inability to directly access the world beyond human reason meant that questions of ontology were effectively ruled out of discussion. For both Immanuel Kant and Adam Smith, key thinkers and founders of the eighteenth-century Enlightenment tradition, the world beyond human access was unproblematic. What did not appear transparently to us, what was not available to our understanding and to our control was therefore of little immediate scientific concern. The approach of making the world transparent, through processes of abstraction, representation and homogenisation, was essential to modernist forms of knowledge and governance, separating a sphere of the known from the unknown, violently forcing the bifurcation of the world between the seen and the unseen, captured well in James C. Scott's classic analysis *Seeing Like a State* (1998; see also Adorno and Horkheimer 1997 [1944]; Foucault 1989 [1966]; Latour 1993).

Hope as the sphere of the unseen – as a space beyond the 'world' of the moderns – was therefore already ontologically articulated at the very heart of

Enlightenment reasoning – not as a temporal or futural beyond but a 'beyond' very much present in the here and now. For Kant's transcendental idealism, this was clearly expressed in the inaccessible gap between appearance and reality, between the phenomena and the noumena. There was a 'beyond' to our knowledge of the world, one closed off to us, but nevertheless available as a secular source of hope that things could always be more perfectible; that this was never all that there was. It is this aspect of potentiality, a potentiality that remains hidden in the present, that gives modernist hope its secular, speculative and futural character.

**Hope in the Anthropocene**

The Anthropocene brings this ontological framing of hope in modernity to a close. There can be no assumption that life has an immanent drive that is 'providential' to humanity or that Enlightenment understandings of progress can be seen to be working in line with the deeper needs of the planetary Earth. One of the key aspects of the Anthropocene is that the divide between the transparent world, understood to be available to instrumental reason, and the unseen or unknowable is no longer so straightforward. For Immanuel Kant and Adam Smith the borders of the inside and outside were necessarily clear, for example, in the divide between appearance and essence or between the sphere of state regulation and market freedoms. In the Anthropocene, understood as a new era in which humankind has fundamentally impacted the Earth systems of environmental and climatic stabilisation, the divide between the known and the unknown has been problematised. With this problematisation, modernist imaginaries of progress and universal causality have been thrown into question and aspects of uncertainty and unpredictability have come to the fore. This is in part because the Anthropocene is seen to have emerged behind the backs of political reason, unseen and unintended (Chakrabarty 2021). Timothy Morton (2013) refers to this as the age of Hyperobjects, entities that we cannot grasp as we are entangled with them rather than observing them from afar.

Today, the view of progress as the building up of a universal store of knowledge of a fixed or determined world of causality no longer appears possible. There is no 'pure' world of nature external to the human sphere of direction and control. The end of this division was already pre-empted in understandings of globalisation as ushering in a new world of risk and uncertainty. In this world governments were no longer initiating policy, acting in a fixed world amenable to instrumental understandings, but were governing 'recursively', responding to the unseen and unintended consequences of previous policy interventions (see, for example, Beck 1992; Giddens 1999). The unseen gradually seeped into the world of governance and policymaking, forcing a new sphere of thought and practice onto the policy agenda. The difference with modernist assumptions of the unseen was that in this 'second modernity' or 'runaway' world of 'global risk society', the unseen was understood to have potentially catastrophic effects.

The passive or optimistic approach to unseen forces and interactive effects, as articulated in modernity, is today likely to be understood as an ideologically charged discourse of denial in the face of catastrophic climate change (Chandler 2019). Modern hope was reliant upon on a linear telos of progress and an underlying assumption of harmony between the world of modernist reason and the world 'unseen' and inaccessible to us but upon which we are dependent. Hope as constructed within modernity today stands accused not just of climate change denial but also of brushing under the carpet all the deaths and dispossessions that were considered necessary for modernity to progress (Colebrook 2020; Povinelli 2021). Discourses of hope in progress and harmony (despite the disastrousness of appearances) are seen to be dangerously ideological: they legitimise history from the perspective of power and subordination.

As Elizabeth Povinelli notes, it is important 'to remember the function of the horizon and frontier in liberalism as a mechanism of disavowal' (2021, 38): 'The horizon is liberalism's governmental imaginary, its means of bracketing all forms of violence as merely unintended, accidental, and unfortunate consequences of liberal democratic unfolding' (2021, 41). It should be highlighted here that the ideological nature of modernist hope, inseparable from the liberal telos of progress, is brought to the fore in relation not only to environmental and species destruction but also the disavowal of modernity's imbrication within chattel slavery, Indigenous dispossession, colonialism and racial capitalism. Hope, in liberal promises of freedom, emancipation and social equality, is increasingly condemned for its structural reproduction of these conditions.

As Rinaldo Walcott argues, in *The Long Emancipation* (2021): 'all of our present conceptions of freedom, understood within that linear progressive narrative, actually prohibit Black subjects' access to that very same linear modernist freedom' (2021, 3). Hope is thus enrolled in processes of tutelage, apprenticeship and subordination, legitimising existing hierarchies, inequalities and exclusions, while dominant liberal universalist understandings are reinforced rather than challenged (see also Lindroth and Sinevaara-Niskanen 2022). As Jovan Scott Lewis explores, even in postcolonial framings of colonial debt and reparation, hierarchies and dependencies are reproduced and affirmed (2020, 150–4; see also Robinson 2020, 226 on 'benevolent violence'). In the Anthropocene, hope can no longer be dependent upon linear temporalities of progress or providential certainties of a 'happy ending'.

As stated above, in the Anthropocene we realise that the unseen or unacknowledged reality of the world exceeds the grasp of modernist, Eurocentric and anthropocentric forms of 'reason' and that therefore we need a different way of accessing or knowing the world. We need to 'see' or 'sense' beyond the limits of the modern ontology; beyond what is sometimes understood as the confines of Kantian 'correlationism' (Meillassoux 2008) where all we have are the phenomena of appearances, the world as always already given to thought. Thus, the crisis of modernity reopens metaphysical questions of the human and the world previously bracketed off from ethical and political discourse. For this reason, the return of hope to the centre of ethical and political concern offers an important field for analysis of contemporary social and political theorising. While hope helps to bridge or to move beyond the binary assumptions of the human/nature or subject/object divide, there appears to be little consensus on how hope is put to work after modernity. This chapter therefore seeks to set out an initial heuristic framework for the analysis of contemporary forms of hope. The following sections therefore briefly outline three potential ways of engaging this 'beyond': the first puts the emphasis on access via the speculative imagination, the second stresses the importance of grounded practice and experimentation, and the third reverses the problematic, seeing the 'world' from the figurative positionality of the unseen, from behind the 'veil' of modernity.

## Speculative Hope: Seeing the Unseen

Continuing the speculative trajectory of theorists like Erich Fromm (1968) and Ernst Bloch (1986) associated with the Frankfurt School, contemporary approaches of speculative transcendental hope focus upon moving beyond the world of human exceptionalism and the human/nature divide to develop 'greener' and more 'ecological' sensibilities (Bennett 2010, xiv, 10). For speculative thinkers, we can do this by working on our own powers of becoming affected, our experiential being in the world, cultivating openness to alterity. As Jane Bennett states: 'The ethical task at hand here is to cultivate the ability to discern nonhuman vitality, to become perceptually open to it' through techniques of 'training oneself' (2010, 14). Good examples of advocates of speculative hope in international relations are William Connolly (2011; 2013) and Jairus Grove (2019). Drawing on Connolly, Grove rejects the 'cryptoprovidence' of Kant (2019, 234) and asserts that hope lies in understandings of complexity and creativity that are 'nonprovidential' (2019, 239). Hope, for Grove and for Connolly (as for Bennett), is a speculative practice oriented towards the virtual world beyond that of actualized appearance.

Thus, 'Connolly's political theorist as seer . . . attempts to peer into the future, but the seer looks for incipient possibilities, not catastrophic certainties' (Grove 2019, 264). Drawing on Deleuze, Connolly argues for a preemptive generosity and openness to the inaccessible beyond, imagining 'a seer dwelling within a nest of partly formed potentialities jostling against and upon each other during a forking moment, with no potentiality settled enough to be foreseen with certainty' (Connolly 2011, 158). Rather than focusing on saving the world as it exists in appearance, speculative political theory, 'the politics of becoming', works within the ontology of hope, one that is affirmative and 'restores belief in the world' (Grove 2019, 270) and 'suggests an enhanced attentiveness to materiality and the chaos of becoming' (Grove 2019, 269). As Connolly suggests, 'A seer by definition lives at the edge of power and events' (2011, 159), living not at the centre of a modernist imaginary, but 'dwelling sensitively' (2011, 165), on the edge, with incipient hope in the world's becoming.

Speculative hope, seeking to free thought from the confines of the linear temporality of modernity, inculcates a practice of openness to the other, to the unseen and to the inaccessible beyond, aware of the subject's mutual imbrication within 'the world of becoming'. Existing within modernity but focused on modernity's unseen, speculative hope occupies a liminal realm between being and non-being or the actual and the virtual. The speculative imaginary is one of the subject staring into the unknown abyss but without fear, instead with an approach of openness and welcome, affirming the world beyond rationalist appropriation. Increasingly, what is valued is otherness in-itself. As Grove states, the end of the world 'is the end of something but never *the* end' (2019, 280, italics in original). It is the lack of certainty that is the source of hope rather than expectations of Kantian providentiality. As Grove concludes his book *Savage Ecology*: 'I am experimenting with the role of seer in order to push further into the metaphysical fallout of cosmic fragility' (2019, 280).

The subject-centred and idealist perspective of speculative hope can also be seen at work within Afrofuturist approaches which focus upon the transcendental subject's inner powers of thinking beyond linear time and space. The 'politics of becoming' and 'unscripted encounter' are vital to the project of escaping the 'linear progress narrative' with its 'Middle Passage epistemology' of continuity, identity and struggle against oppression, with ontological assumptions of fixed origins, laws of cause and effect and collective representations (Wright 2015, 26). Instead, Afrofuturist approaches turn to non-linear models of spacetime, focusing upon the present as a moment of openness and emergence – what Michelle Wright calls the 'epiphenomenal time' of the 'encounter', where the individual intersects in the now with a multitude of possible futures (2015, 31). Wright argues that epiphenomenal time restores agency to the subject:

> Agency here is not tied to concrete outcomes (born of concrete goals) but to the choice to notice and wonder at differences that the linear progress narrative struggles to wholly interpellate on its own . . . interpellation begins with the self, it is not a reactive action but one of 'choice'. (Wright 2015, 117)

Wright's *Physics of Blackness* has been influential for the Black Quantum Futurism collective, which draws on quantum understandings along with African cosmologies to train and develop capacities for reaching into the unknowable beyond through the extension of access in the present (Phillips 2015; see also Gordon in Chapter 9), 'decolonizing the mind' so that the past, present and future become not separate entities but overlapping dimensions (VerCetty 2020, 140). The speculative beyond provides the fundamental break required from what some see as 'time-warped trauma', 'the stuckness' in the horrors of the Middle Passage and chattel slavery (Womack 2016, 59).

For many Afrofuturists, the positionality of the 'seer', in modernity but staring into the beyond, is an unrequested gift of societal exclusion. Jayna Brown writes:

> I argue that being categorised as inhuman, or not quite human, is a privileged position from which to undo the assumptions not only of race thinking but of the other systems of domination . . . and instead marvel at the potential modes of existing as biological entities such exclusion opens up. (2021, 112)

For Brown, this turn to the speculative outside of the ontology of being is necessarily 'a jump into the unknowable . . . not the unknown' (2021, 6–7). '[B]ecause black people have been excluded from the category human, we have a particular epistemic and ontological mobility . . . we develop marvelous modes of being in and perceiving the universe' (2021, 7). She ends *Black Utopias* on a similar note to Grove, with an analysis of Alex Garland's film *Annihilation*, where the sole black woman, trapped in the mysterious Area X, accepts the abyss, becoming a flowering bush, fulfilling 'a radical longing to merge with the cosmos' and to 'join the awesome, the unexpected, *already present* in the world' (2021, 178, italics added).

Speculative hope thus can be seen as part of a critical tradition that consciously breaks from the linear and providential understanding of Kantian hope in modernity. However, the break from Kantianism can be questioned to the extent that there is still a reliance on the transcendental subject breaking with the modern ontology or 'decolonizing' its mind. In order to be open to the present, minds need to be 'freed' so that the unseen potentiality existing in the world can be enabled.

**Pragmatic Hope: Worlding Worlds Otherwise**

Pragmatic hope is less about changing modes of thought and more about our embodied and embedded modes of practice. For pragmatic hope the worlds we live in are not transcendental products of our minds but of our material being as worldmaking agential beings entangled with others. Pragmatic hope lacks the abstract framings of speculative hope, where there is the world of the known and the unknowable, of the actual and the virtual, of modern ontology and the unseen abyssal 'beyond'. Indigenous theorist Deborah Bird Rose (2011) argues that 'worlds' are practices, or modes of entangled being. How we world the world has consequences; our being in the world can contribute to collective flourishing or collective disaster. As we are aware, in the Anthropocene, the modern mode of being in the world is reproducing cycles of death and destruction rather than enabling processes of growth and differentiation. While Indigenous peoples are held to cultivate their environments in mutual life-giving ways, the moderns have instigated feedback loops of 'deathwork', wiping out ecological diversity and threatening mass species extinction. Rather than affirming the unknowable possibilities of the present, pragmatic hope suggests an attentiveness to the links, relations and connections in the here and now but beyond the reach of human cognition. Often this requires attentiveness to the knowledge and understanding of patterns and correlations, often involving the actions of nonhumans, to access and amplify the power of life as a self-organising system.

I would argue that pragmatic hope is to modern hope what neoliberalism is to liberalism. While for Kant the power of life or nature was unknowable but necessarily assumed to be providential, for pragmatic hope the relational forces of life need to be channelled in ways that amplify their negentropic powers. This distinction mirrors that between free-market liberalism, where the 'invisible hand' of the market was understood to deliver the best possible outcomes, and neoliberal or neo-institutionalist approaches which sought to indirectly shape market systems (Hayek 1960; North 1990). Neoliberal approaches differed in their awareness that markets' immanent, relational, power was not 'naturally' self-organised for providential outcomes but could also result in catastrophic collapse or system change. Markets needed to be governed for rather than directed or left to laissez-faire (Foucault 2008, 131). For pragmatic hope, then, the question is how to materially detect the signs, effects or registrations of these vital forces whose operation is beyond human powers of direct access and understanding. Thus, pragmatic hope moves beyond modernist ontology in an imaginary which is often conceived as 'after modernity', after the naive or hubristic assumptions of the naturalness of 'nature' or of markets. Perhaps the archetypal figure of pragmatic hope is Bruno Latour, with his assertion that *We Have Never Been Modern* (1993). For Latour, we have always worked pragmatically although we told ourselves fictions that we were discovering some objective or universal laws of cause and effect.

Pragmatic hope seeks to amplify and to tap into the powers of relational interaction, the powers of life and of markets, indirectly, not through top-down processes of control and direction, but through the bottom-up tracing of effects. As already intimated above, in the work of Bird Rose, there is a close relation to Indigenous understandings of multi-perspectivism, which also engage with the search for patterns and correlations, described as *Cannibal Metaphysics* by Eduardo Viveiros de Castro (2014) and as 'material semiotics' by Eduardo Kohn (2013). The search for patterns and correlations, to access the world beyond the reductionist imaginary and linear causality of modern ontology, involves tracing the feedback loops and sensing processes of emergence. Giving 'what is' its due enables an attention to Big Data and other forms of data gathering and sensing for correlations in more-than-human assemblages. A good recent example of pragmatic hope would be Benjamin Bratton's *The Revenge of the Real: Politics for a Post-pandemic World* (2021) which seeks to develop new forms of planetary governance based upon a positive biopolitics of sensing and an understanding of humans as material, biological and epidemiological agents rather than as rational actors. Bratton argues that rather than working with fixed categories of representation, mobile, fluid approaches of tracking and tracing enable 'a politics capable of engagement with the full complexity of reality' (2021, 14).

At the more radical end of approaches that work through pragmatic hope, some radical Indigenous and critical feminist theorists argue that Eurocentric versions of pragmatic tracing and governance are problematic. This is due to the fact that they seek to generalise a certain set of techniques as if there were some timeless understandings or methods available to access the unseen world beyond modernist appropriation. These (Eurocentric) approaches to hope, according to Povinelli (2021, 126), seek to make hope 'compatible with liberal and illiberal capitalism'. Thus, relational understandings, often appropriated from Indigenous communities, are instrumentalised as liberal powers 'want to continue to have what they have'; 'What they want is for Indigenous people to save the world' (Povinelli 2021, 126; see also Chandler and Reid 2019). Povinelli seeks to challenge the sometimes abstract and timeless ontological understandings of both speculative and pragmatic approaches to hope. Instead of starting from metaphysical claims, she locates hope in the pluralised grounded practices of resistance to the forces of colonial modernity and racial capitalism. In so doing, she articulates a different spatial and temporal framing of pragmatic hope, one that is horizontal rather than vertical, critical rather than affirmative.

In Povinelli's articulation of pragmatic hope there is a critical shift in positionality away from hegemonic Eurocentric framings. She makes the point that hope is not about 'survivance' in the wake of a coming environmental catastrophe. Instead, she highlights what she calls 'the ongoing nature of the ancestral catastrophe of colonialism and its epistemological and ontological presuppositions and unfoldings' (2021, 132). Pragmatic hope is no longer about saving the liberal 'world of modernity' from impending crisis, as it is for Bratton and Latour, but rather about mobilising 'spatial and affective discourses in order to transform actual harms into horizontal hopes' (2021, 132). It is about slowing and challenging the ongoing sacrificing of worlds called for in order to salvage the world of liberal modernity. Povinelli is careful to articulate the necessity of grounded practices which cannot be universalised or instrumentalised by others. This attention to multiplicities, which is grounded in Marilyn Strathern's anthropology of *Partial Connections* (2004), has also been an important influence upon approaches to pragmatic hope as presented in Haraway's *Staying with the Trouble* (2016) and Anna Tsing's 'life in the ruins' (2015).

**Nihilist Hope: Theorising from the Abyss**

The third framing, nihilist hope, could also be seen as existing in a field of conceptual work having both a Eurocentric and a more critical contemporary framing; in this case, the more radical expressions could be seen as those of critical Black studies and work associated with Afropessimism. Earlier versions, associated with Walter Benjamin and Theodore Adorno of the Frankfurt School, find hope in the radical negation of the world of unending violence, where both carrying on in the world and giving up in despair are impossible. Adorno states that the only hopeful position is that of nihilism, the hope of negation rather than the abstract desire for nothingness. Drawing upon Beckett, he argues:

> . . . the created world is radically evil, and its negation is the chance of another world that is not yet. As long as the world is as it is, all pictures of

reconciliation, peace, and quiet resemble the picture of death. The slightest difference between nothingness and coming to rest would be the haven of hope, the no man's land between the border posts of being and nothingness . . . The true nihilists are the ones who oppose nihilism with their more and more faded positivities, the ones who are thus conspiring with all extant malice, and eventually with the destructive principle itself. Thought honours itself by defending what is damned as nihilism. (2007 [1966], 381)

For Adorno, the dark side of modernist 'reason' is the world as concentration camp. 'A new categorical imperative has been imposed by Hitler upon unfree mankind: to arrange their thoughts and actions so that Auschwitz will not repeat itself . . . The course of history forces materialism upon metaphysics, traditionally the direct antithesis of materialism' (2007, 365). It might, at first glance, appear that nihilism and hope are incompatible concepts; however, this is far from the case. It is the force of critique of what exists that creates the possibility of another world to come after (see also Waldow in Chapter 11). From the perspective of nihilist hope, the critical approaches of pragmatic hope that insist on 'staying with the trouble' and enabling 'life in the ruins' are still problematic and affirm the world that exists through, as Adorno says, 'their more and more faded positivities'. As Claire Colebrook notes, pragmatic approaches to hope, even the more radical ones described above, as articulated by Donna Haraway and Anna Tsing, also have 'a practical and therapeutic function, enabling us to continue to be who we are' (2020, 185); an 'ongoing managerial exercise in anthropodicy: what we have been in the past may have been destructive, but future non-being is unthinkable' (2020, 186).

Nihilist hope seeks to move beyond the privileging of modernist ontology as the starting point, but without affirming the world as it exists. It is not just the modernist construction of the 'human' but the ontology of 'world' itself that is problematised in the nihilist understanding of the violence at the heart of modernist cuts and binaries. As Calvin Warren states, 'blackness is outside ontology' understood as grounded in 'an antiblack metaphysics' (2018, 42–3). Where nihilist hope differs from speculative and pragmatic approaches to hope is in locating a positionality *outside* the world as grasped in modernist constructions of the human as subject. The starting point is a positionality understood as 'within the veil' (Du Bois 2018 [1920]), external to the 'world' understood as a world made and sustained by the violence of ontological terror. Nihilist hope depends upon the inversing or negating of the modernist framing of the unseen, theorising *from* rather than *towards* the 'abyss' (see Pugh and Chandler 2023). This positionality is that of the 'other' aware of its 'otherness', often expressed in W. E. B. Du Bois's concept of 'double consciousness' (1903). As

Paul Gilroy states, for critical Black studies Du Bois leads the way in theorising modernity via a 'sustained and uncompromising interrogation of the concept of progress from the standpoint of the slave' (1993, 113).

One of the most powerful expositions of nihilist hope is Du Bois's short story 'The Comet' (1920). In the story, a Black worker is forced down into a New York building's vaults to undertake work 'too dangerous for more valuable men'. When he emerges, a comet has passed close to the Earth seemingly emitting deadly gases which have killed everyone on the surface. Coming to terms with life after the ending of the world, the man falls in love with a white woman who has also survived, a relationship that would have been impossible otherwise. Swept up in their emotions, the ending of the world appears as positive:

> 'Death, the leveler!' he muttered.
> 'And the revealer,' she whispered gently, . . .

The ending of the world is a moment of emancipation from the psychological and material 'shackles' of racial division, but also is 'revealing' of the unseen human potential that is routinely disavowed (for an exploration of this dialectic in Du Bois's visual statistical work, see Phull in Chapter 4). The sad ending of the story is that the destruction is only localised to New York and the normality of racial domination is quickly restored as people return: the man is threatened with lynching after being spotted with the white woman. For nihilist hope, the unseen world 'behind the veil' requires as its precondition the ending of the world of modernity. Advocates of nihilist hope are thereby drawn to the world-ending stance of theorists working in critical Black studies, particularly articulating their work in alignment with Afropessimism. As Colebrook states:

> This is what I take Afro-pessimism's conception of social death to be, an awareness not so much that one does not have a world or belong in the world, but that the world demands one's non-being. Currently this form of existence is utterly tragic, constantly resulting tracing the wake of black lives not mattering. Even so, Afro-pessimism also offers a positive sense of the end of the world, where non-being and worldlessness provoke thought to move beyond the world. (2020, 197)

Nihilist hope does not provide a speculative imaginary that presumptively welcomes the unseen 'beyond', nor does it provide an affirmation of forces and effects that exist in potentiality 'here and now'. Nihilist hope is the hope of negation, facilitating the practical tasks of deconstruction, of paraontological unravelling (on the concept of paraontology, see Chandler 2014).

## Conclusion

Questions of ontology, of the 'unseen' beyond modernist reason – the unintended feedback effects, the capitalist 'externalities', exclusions and disavowals from environmental destruction to hidden pasts of chattel slavery and racial capitalism – are increasingly central to social and political thought in the epoch of the Anthropocene. This shift has put hope at the forefront of contemporary concerns, but not the hope of modernity with its confidence in a 'happy ending' no matter how many setbacks are experienced. Hope in the Anthropocene conceives itself to be less 'human-centred', operating beyond the constraints of the human/nature divide. The hope of the speculative 'seer' is in a beyond that should be welcomed with presumptive generosity and non-anthropocentric care, rather than suborning ourselves to what exists. Pragmatic hope, in the assumption that modernity is over, is affirmative of non-anthropocentric worldmaking practices, engaging in emergent multiplicities in ways that seek to inculcate differentiation rather than to direct and control. Nihilist hope, instead, confronts modernity from its constitutive 'beyond', engaging in the work of paraontological disruption and deconstruction, placing hope in negation.

## Acknowledgement

This chapter draws on material previously published: 'The Politics of the Unseen: Speculative, Pragmatic and Nihilist Hope in the Anthropocene', *Distinktion: Journal of Social Theory* (published online 19 July 2023).

## References

Adorno, T. 2007 [1966]. *Negative Dialectics*. New York: Continuum.
Adorno, T. and Horkheimer, M. 1997 [1944]. *Dialectic of Enlightenment*. London: Verso.
Beck, U. 1992. *Risk Society: Towards a New Modernity*. London: Sage.
Bennett, J. 2010. *Vibrant Matter: A Political Ecology of Things*. Durham, NC: Duke University Press.
Bird Rose, D. 2011. *Wild Dog Dreaming: Love and Extinction*. Charlottesville: University of Virginia Press.
Bloch, E. 1986. *The Principle of Hope*, vol. 1. Cambridge, MA: MIT Press.

Bratton, B. 2021. *The Revenge of the Real: Politics for a Post-pandemic World*. London: Verso.
Brown, J. 2021. *Black Utopias: Speculative Life and the Music of Other Worlds*. Durham, NC: Duke University Press.
Chakrabarty, D. 2021. *The Climate of History in a Planetary Age*. Chicago, IL: University of Chicago Press.
Chandler, D. 2019. The death of hope? Affirmation in the Anthropocene. *Globalizations* 16(5): 695–706.
Chandler, D. and Reid, J. 2019. *Becoming Indigenous: Governing Imaginaries in the Anthropocene*. London: Rowman & Littlefield.
Chandler, N. D. 2014. *X: The Problem of the Negro as a Problem for Thought*. New York: Fordham University Press.
Colebrook, C. 2020. What would you do (and who would you kill) in order to save the world? Dialectical resilience. In D. Chandler, K. Grove and S. Wakefield (eds), *Resilience in the Anthropocene: Governance and Politics at the End of the World*. Abingdon: Routledge, 179–99.
Connolly, W. E. 2011. *A World of Becoming*. Durham, NC: Duke University Press.
Connolly, W. E. 2013. *The Fragility of Things: Self-Organizing Processes, Neoliberal Fantasies, and Democratic Activism*. Durham, NC: Duke University Press.
Du Bois, W. E. B. 1903. *The Souls of Black Folk*. Project Gutenberg. Available at https://www.gutenberg.org/files/408/408-h/408-h.htm (accessed 21 December 2023).
Du Bois, W. E. B. 1920. The Comet. Available at http://zacharyrawe.com/sem_6_the_comet_dubois.pdf (accessed 21 December 2023).
Du Bois, W. E. B. 2018 [1920]. *Darkwater: Voices from Within the Veil*. CreateSpace Independent Publishing Platform.
Foucault, M. 1989 [1966]. *The Order of Things*. Abingdon: Routledge.
Foucault, M. 2008. *The Birth of Biopolitics: Lectures at the Collège de France, 1978–1979*. Basingstoke: Palgrave Macmillan.
Fromm, E. 1968. *The Revolution of Hope: Toward a Humanized Technology*. New York: Harper & Row.
Giddens, A. 1999. *Runaway World: How Globalization is Reshaping Our Lives*. London: Profile Books.
Gilroy, P. 1993. *The Black Atlantic: Modernity and Double Consciousness*. London: Verso.
Grove, J. 2019. *Savage Ecology: War and Geopolitics at the End of the World*. Durham, NC: Duke University Press.
Haraway, D. 2016. *Staying with the Trouble: Making Kin in the Chthulucene*. Durham, NC: Duke University Press.

Hayek, F. A. 1960. *The Constitution of Liberty*. Abingdon: Routledge.
Kant, I. 1991. Idea for a universal history with a cosmopolitan purpose. In I. Kant, *Political Writings*. Cambridge: Cambridge University Press, 41–53.
Kohn, E. 2013. *How Forests Think: Toward an Anthropology beyond the Human*. Berkeley, CA: University of Calfornia Press.
Latour, B. 1993. *We Have Never Been Modern*. Cambridge, MA: Harvard University Press.
Latour, B. 2021 *After Lockdown: A Metamorphosis*. Cambridge: Polity Press.
Lewis, J. S. 2020. *Scammer's Yard: The Crime of Black Repair in Jamaica*. Minneapolis: University of Minnesota Press.
Lindroth, M. and Sinevaara-Niskanen, H. 2022. *The Colonial Politics of Hope: Critical Junctures of Indigenous-State Relations*. Abingdon: Routledge.
Meillassoux, Q. 2008. *After Finitude: An Essay on the Necessity of Contingency*. London: Continuum.
Morton, T. 2013. *Hyperobjects: Philosophy and Ecology after the End of the World*. Minneapolis: University of Minnesota Press.
North, D. C. 1990. *Institutions, Institutional Change and Economic Performance*. Cambridge: Cambridge University Press.
Phillips, R. (ed.) 2015. *Black Quantum Futurism: Theory and Practice*, vol. 1. Philadelphia, PA: AfroFuturist Affair/House of Future Sciences Books.
Povinelli, E. 2021. *Between Gaia and Ground: Four Axioms of Existence and the Ancestral Catastrophe of Late Liberalism*. Durham, NC: Duke University Press.
Pugh, J. and Chandler, D. 2023. *The World as Abyss: The Caribbean and Critical Thought in the Anthropocene*. London: University of Westminster Press.
Robinson, D. 2020. *Hungry Listening: Resonant Theory for Indigenous Sound Studies*. Minneapolis: University of Minnesota Press.
Scott, J. C. 1998. *Seeing Like a State: How Certain Schemes to Improve the Human Condition have Failed*. New Haven, CT: Yale University Press.
Smith, A. 2022 [1776]. *Wealth of Nations, Books I–III*. Ottawa, Canada: East India Publishing Company.
Strathern, M. 2004. *Partial Connections*. Lanham, MD: Rowman & Littlefield.
Tsing, A. L. 2015. *The Mushroom at the End of the World: On the Possibility of Life in Capitalist Ruins*. Princeton, NJ: Princeton University Press.
VerCetty, Q. 2020. Building the X variable: Artivism with Afrofuturism. In R. Phillips (ed.), *Space-Time Collapse II: Community Futurisms*. Philadelphia, PA: AfroFuturist Affair/House of Future Sciences Books, 133–51.
Viveiros de Castro, E. 2014. *Cannibal Metaphysics*. Minneapolis, MN: Univocal.
Walcott, R. 2021. *The Long Emancipation: Moving Toward Black Freedom*. Durham, NC: Duke University Press.

Warren, C. 2018. *Ontological Terror: Blackness, Nihilism and Emancipation*. Durham, NC: Duke University Press.

Womack, Y. 2016. Thoughts on navigating the time warp of horrors and riding the DNA strands of resilience. In R. Phillips (ed.), *Space-Time Collapse I: From the Congo to the Carolinas*. Philadelphia, PA: AfroFuturist Affair/House of Future Sciences Books, 55–64.

Wright, M. E. 2015. *Physics of Blackness: Beyond the Middle Passage Epistemology*. Minneapolis: University of Minnesota Press.

# 2

# A Hope Against Hope: Scandal, Cynicism and Critique in the Wake of the COVID-19 Polycrisis

*Chris Zebrowski*

**Introduction**

Crises, by definition, are a time of significant potential for change (Roitman 2014). Crises signify a moment of interruption or breakdown in a given order. In compromising an order, crises suspend habituated ways of acting in the world, forcing us to reconsider our place within it. Such disruption compels us to search out causes which can, in turn, expose structural problems, unknown or disregarded, within a given order. Crises are thus critical moments in relation to which significant changes can be fostered.

Hope arises from a conviction in the possibility of change. Hopes (and fears) may be invested within different figures of crisis (past/future/present) as critical opportunities for change, just as critique has long hoped to exercise the power of instigating crises to incite change (Koselleck 2006). However, hope can also be problematic. One can harbour *false* hopes, allow fantasies to become a *barrier* to productive change, and one can *lose* hope.

So, what are the conditions for hope in the wake of COVID-19? Some scholars see, in the devastation wrought by the pandemic, reasons to be optimistic. Gross writes that COVID-19 has 'loosened neoliberalism's hegemonic grip on the future' (2022, 448) and opened a discussion on how we might build back better. In upsetting the hegemony of the current order, others see an opportunity for the crisis to advance alternative governmental systems, more capable of managing the crises of the future: social democracy (Walby 2021) or communism (Žižek 2020). Still others have written about the potential for lessons to be learned from the experience of COVID-19 that might be useful in compelling meaningful action to address the existential problem of global climate change (Cole and Dodds 2021; Latour 2020). For many scholars, hope in the wake of COVID-19 is linked to its capacity to deliver transformational changes to the systems of governance ordering our lives.

I certainly hope that some good will come of our collective experience of COVID-19 as well. However, at least at the time of writing, the experience of the pandemic has failed to instigate any widespread rethink of the prevailing economic and political order (Šumonja 2020). In Britain, the governmental rationality of neoliberalism appears to have been wholly unaffected by the crisis as each of the major political parties prepares the public for renewed economic austerity. It is certainly true that critical reflection on the lessons learned from the COVID-19 pandemic have been crowded out in the broadcast media by a new set of attention-demanding crises: the inflation crisis, the migration crisis and culture wars (amongst numerous others). But distraction only explains part of it. Clearly, there is a significant portion of the population who do not desire change, who despite the devastation wrought by the COVID-19 pandemic yearn only for things to return to normal (Briggs et al. 2020).

To hope for a return of calm and predictability from within a period of crisis is understandable enough. Yet it does raise the question as to why a 'return to normality' has such appeal within societies which COVID-19 revealed to be afflicted by staggering levels of inequality, precarity and insecurity.

This chapter aims to interrogate the problematic figure of hope as it has been enacted in the wake of the COVID-19 crisis. I reflect on the complex, if not contradictory, ways hope and hopelessness are manifest within this period of multiple overlapping crises, asking whether we are experiencing a crisis of hope, and what can be hoped for in a world so afflicted by crisis. It begins by situating the (recently fashionable) concept of 'polycrisis' alongside Reinhart Koselleck's etymology of crisis to raise questions concerning the historical significance of this new mode of understanding and experiencing crisis. The chapter then moves on to consider the conditions for hope in a time of polycrisis by considering how critique, cynicism and cruel optimism were enacted within the British media. It will conclude with a reflection on the role of critique and the possibility of holding out a 'hope against hope' in an era of polycrisis.

### The Time of Polycrisis

In the new normality that is taking shape in the wake of the COVID-19 pandemic, the concept of 'polycrisis' is gaining traction. Said to have been coined by French theorist of complexity Edgar Morin and co-author Anne Brigitte Kern (1999), the term was recovered by former European Commission president Jean-Claude Juncker (2016) and then popularised by the economic historian

Adam Tooze (2022). Polycrisis refers to the endogenous, self-amplifying crises that emerge within complex systems. The significance of the polycrisis idea is that it refers not only to the number and diversity of contemporary crises, but to the complex entanglement and interaction of these distinct crises within a singular polycrisis. These complex interactions in turn give rise to emergent phenomena, 'so that the whole is even more overwhelming than the sum of the parts' (Tooze 2022).

COVID-19, in this respect, might be regarded as a paradigmatic example of a polycrisis (Tooze 2021). The COVID-19 polycrisis was experienced as a multiplicity of diverse yet interrelated crises: a public health crisis *and* a political crisis *and* an economic crisis, and so on. Triggered by the release of a novel coronavirus crossing species lines, the COVID-19 polycrisis advanced by connecting with and exploiting a series of endemic crises – racism, environmental destruction, economic inequality – that have long waged a slow and attritional violence (Anderson et al. 2020; Nixon 2011), but which within this new configuration realised a new lethality. The complex, distributed and unpredictable emergence of the COVID-19 polycrisis is thus exemplary of the dynamic multiplicity of entangled crises anticipated to characterise this epoch of the Anthropocene (Chakrabarty 2018; Chandler et al. 2020; Tooze 2021).

While some have questioned whether 'polycrises' are in fact novel phenomena, or simply 'just history happening' (Drezner 2023), I am more interested in what the ubiquity of this term says about our general understanding and experience of the world at this present conjuncture. Polycrisis articulates a new imaginary of danger. It speaks to a *feeling* of insecurity: an acute and pervading sense of lingering exposure to catastrophic threats. Reports that 'polycrisis' was the most 'ubiquitous buzzword of this [2022] year's Davos gathering' (Serhan 2023) attest to the wide breadth of this sentiment.

Polycrisis might be understood as the latest development within the long evolution of the concept of crisis. In his masterful conceptual history of crisis, Koselleck (2006) traced the etymological roots of the concept of crisis to the Ancient Greek *krinô*, meaning to separate, to choose, to decide and to judge. The term, Koselleck tells us, had distinct, yet related, applications in the fields of medicine, law and Christian theology. In medical discourses, crisis referred to a critical time in which an urgent medical intervention was required to ensure survival (Koselleck 2006). Crises demanded a diagnosis – a judgement as to the cause of illness – as well as an appropriate intervention. In legal circles, the moment of crisis similarly referred to the judgement, trial and decision of the court. Theological references to crisis appear to combine the medical and juridical senses of this term. Crisis, within Christian theology, became the moment of God's judgement and intervention. This act of judgement is, of course, linked to the Christian promise of salvation, and the concept of crisis gained critical purchase within apocalyptic narratives of the Last Judgement. It was here, Koselleck stresses, that the idea of crisis took on a new sense – no longer simply referring to a critical time of decision but placed at the terminus of a teleological account of history: the moment of crisis 'at the end of the world [that] will for the first time reveal true justice' (Koselleck 2006, 359).

Truth, diagnosis and revelation are at the heart of *krísis*. The notion of *krísis* is itself an adaptation of the earlier Proto-Indo-European root *\*krei-*, meaning 'to sieve, discriminate, distinguish' and which is the root of contemporary concepts of certainty: for example, ascertain, discern, discriminate. *Krísis* refers to a time in which a truth is revealed, a judgment is made and an action is taken (Roitman 2014). It is in this sense that the concept of crisis is intimately linked to critique. The root *\*krei* is etymologically shared by both the concepts of crisis and critique.

Koselleck (2006) traced the elliptical return of crisis and critique from their shared etymological origins to their conceptual re-affiliation in Europe between 1780 and 1850. According to him, these terms became re-linked within scholarly discourses once 'history' and 'society' were discovered as secularised realms of human action and autonomy. 'Applied to history', Koselleck writes, '"crisis," since 1780, has become an expression of a new sense of time which both indicated and intensified the end of an epoch' (2006, 358). Here, crisis takes on an epochal meaning, as a critical transition point in world history. Major historical events, such as the French Revolution, displaced the Christian apocalypse as crises that gave direction to history and propelled it forward. In turn, the hope of Christian salvation is sublimated into the secularised hopes that such crises will yield Enlightenment progress. The elevation and secularisation of crises, as events comprising the engine of history, placed a new emphasis on the responsibilities of scholarly critique. Critique was not simply an intellectual practice aiming to discern and reveal the drivers of crises, but an intellectual act which could provoke crises and change world history. As Paul de Man once stated, 'all true criticism occurs in the mode of crisis' (de Man 1967, 44).

For Koselleck, the secularisation of the concept of crisis, and attendant translation of Christian ideas of salvation into an Enlightenment hope in progress, elevated crisis to a key term of art underpinning modernist accounts of history from the late eighteenth century. With its roots in the complexity thinking of Edgar Morin (Morin and Kern 1999), the concept of polycrisis represents the most recent evolution of this ancient concept: the postmodern correlate of the new earthly epoch of the Anthropocene. In distinguishing itself from the 'crises' of modernity, polycrisis would appear to announce a new understanding and experience of time. Given the centrality of the figure of crisis to the temporal understanding and experience of modernity, we should be compelled to inquire what implications the advent of the polycrisis has for our relation to (historical) time.

For our purposes, I would like to reflect on three temporal modifications associated with the notion of polycrisis, before turning to consider its implications for hope. First, we must recognise polycrises as a defining feature of the new epoch of the Anthropocene. However, polycrises do not index moments of transition between stable temporal periods as crises did for modernist historians. The Anthropocene is an epoch *of* polycrises, a period marked by the permanent unfolding of a multiplicity of entangled crises. The multiplicity of entangled crises which define the Anthropocene render problematic any attempt to identify one as the event announcing the advent of, and setting in motion of, the Anthropocene.[1] If crisis is a permanent, enduring condition, then this obviously troubles any sense of normality in distinction to crisis: as a period marked by stability, predictability and safety. Roitman, for one, identifies the oxymoronic sense of this. 'How', she asks, 'did crisis, once a signifier for a critical, decisive moment, come to be construed as a protracted historical and experiential condition' (Roitman 2014)?

Second, we might ask, if crises gave history direction for both moderns and Christians, then what might the effect of polycrisis be for our understanding of the metaphysics of chronological, historical time? For the moderns, crises were exceptional events that propelled historical time forward through stable epochs. Yet, in the Anthropocene the persistence of diverse and entangled crises is a normal, unexceptional condition of this epoch, which, we are told, will be marked by instability. Disaggregated, diverse crises at the molecular level appear contingently, ambivalent to any predetermined historical direction. Yet, at the molar or aggregate level, the combined weight of this multiplicity of crises finds its telos. The unfolding of the polycrisis within the Anthropocene propels history towards a new Apocalypse: the sixth extinction (Kolbert 2014) – massive species death with no clear redemptive function (Colebrook 2014).

The implications of the polycrisis for the understanding and experience of historical time is clear to Morin and Tooze, who both emphasise how the idea of polycrisis breaks from modernist temporalities of world history. Tooze agrees that our experience of the polycrisis unsettles modernist understandings of the teleological direction of history:

> [Polycrisis] registers the unfamiliar diversity of the shocks that are assailing what had previously seemed a settled trajectory of global development. It insists that this coincidence of shocks is not accidental but cumulative and endogenous. And, by its currency, it marks the moment at which bullish

self-confidence about our ability to decipher either the future or recent history has begun to seem at the same time facile and passé. (Tooze 2023)

Morin and Kern (1999) thus associate the polycrisis with the collapse in the 'modern faith' of salvation through development and progress. They insist that the problematisation of the idea of progress, and the attendant promise of salvation, means humanity must (once again) reconcile itself with a pervading sense of doom.

The affective experience of polycrisis is the third temporal modification to consider. If, according to Koselleck, the notion of crisis is firmly rooted in the ideas of diagnosis and decision, then it is significant that Tooze links the idea of polycrisis with disorientation and indecision. According to Tooze, 'What makes the crises of the past 15 years so disorientating is that it no longer seems plausible to point to a single cause and, by implication, a single fix' (Tooze 2022). As Tooze explains, this sense of disorientation is a product of the seeming impossibility of untangling the knotted cluster of contemporary crises to diagnose a singular root cause. The difficulty of distinguishing and diagnosing the multiplicity of polycrises exacerbates indecision despite the urgency of the problems faced. This resonates strongly with new materialist approaches to the Anthropocene which recognise it as more than a geological epoch defined in chronostratigraphic or geochronological terms: the Anthropocene needs to be approached as a predicament that requires a fundamental rethink of how we understand and approach knowledge, ethics, politics, art and society (Haraway 2016; Latour and Porter 2017).[2]

The polycrisis is not a new idea. However, its recent popularisation expresses a widespread desire to get to grips with something specific about the present moment. The idea of polycrisis gives academic explanation to a general feeling. The polycrisis helps to articulate a palpable feeling of precarity, insecurity and disorientation in the wake of our collective experience of COVID-19. It links these feelings to a more general sentiment of anxiety, confusion and powerlessness many feel as we drift towards climate catastrophe. It reflects a sense that the present is markedly different from the past – more dangerous, less sure. And it gives scholarly validation to growing misgivings about the prospects for sustained growth, development and progress. In light of the attention now being afforded to the concept of polycrisis, one might naturally ask about the prospects for hope in the Anthropocene.

## Hope in (Poly)Crisis

Hope arises from a critique of the present and a conviction in the possibility of change. In positing the future as open to difference, hope enacts a present which, though currently unsatisfactory, contains 'something that has not yet realized itself' (Bloch 1996, 193). Hope is fundamentally a temporal relation. Hope anticipates and projects from the present a world that has not-yet become. To hope requires one to believe that things can change and that they should. In doing so, hope cultivates not only an imagined future, but reframes the present as unsatisfactory yet pregnant with possibility. Hope therefore does not simply hinge on the understanding of the possibility of changing the future for the better. It also frames the present as in crisis: 'uncentered, dispersed, plural and partial' (Gibson-Graham 1996, 259). It is in this sense that Bloch insists that 'the world itself, just as it is in a mess, is also in a state of unfinishedness and in experimental process out of that mess' (Bloch 1996, 221). Hope springs forth from a critique of the present and a conviction in the possibility for change. As Marcel suggests, 'the conditions that make it possible to hope are strictly the same as those that make it possible to despair' (Marcel 2007, 101).

But critique does not necessarily lead to hope. In his *Critique of Cynical Reason* Peter Sloterdijk provocatively accuses the Enlightenment of cultivation of a critical mode of thought that culminates in a period of widespread cynicism. For Sloterdijk,

> Cynicism is enlightened false consciousness. It is that modernized, unhappy consciousness, on which enlightenment has laboured both successfully and in vain. It has learned its lessons in enlightenment, but it has not, and probably was not able to, put them into practice. Well-off and miserable at the same time, this consciousness no longer feels affected by any critique of ideology; its falseness is already reflexively buffered. (Sloterdijk 1987, 5)

The Enlightenment, according to Sloterdijk, promised to deliver a universal uplift of humanity ('progress') by developing and applying critical reasoning. All forms of idealism were problematised through an insatiable and all-consuming critical project. The success of this critical project has produced a crisis of critique which he calls cynicism. Sharon Stanley writes that for Sloterdijk cynicism represents 'the final, melancholic resting place of an exhausted critical consciousness' (Stanley 2007, 385). The success of this critical project has made it impossible to posit new worlds in good faith. Those who steadfastly remain committed to progressive political action are held with deep suspicion: at best, charged with being naive and, at worst, charged with hypocrisy. Cynicism arises as a kind of hyperactive critique without hope. Despite finding dissatisfaction everywhere, the cynic loses hope because nothing can ever truly change. The result is a widespread disengagement from politics and a reinvestment in private individualised hopes and dreams. Cynicism shields one from being duped, demonstrating a kind of wisdom that can be used to justify apathy and inaction. It is a defence mechanism for one's ego.

Hope persists in a world marked by rampant cynicism, but it may not be constructive. In her book *Cruel Optimism* (2011), Lauren Berlant draws attention to some of the problematic ways in which hope may manifest:

> A relation of cruel optimism exists when something you desire is actually an obstacle to your flourishing. It might involve food, or a kind of love; it might be a fantasy of the good life, or a political project. It might rest on something simpler, too, like a new habit that promises to induce in you an improved way of being. These kinds of optimistic relation are not inherently cruel. They become cruel only when the object that draws your attachment actively impedes the aim that brought you to it initially. (Berlant 2011, 1)

The concept of cruel optimism is explicitly linked to the decline of the American dream. It interrogates how hope gets contorted when society can no longer provide opportunities for individuals to realise the criteria which that society holds as representative of the 'good life'. According to Berlant, '[i]t is a book about the attrition of a fantasy, a collectively invested form of life, the good life' (Berlant 2011, 11). *Cruel Optimism* is a book not about hopelessness but about the persistence of hope in the face of exhaustion, indifference and disillusionment – hope, perhaps, without optimism. Berlant finds that people still remain attached to fantasies of the good life and invested in a kind of flourishing that is increasingly difficult to attain amid socio-economic conditions of increased flexibility, precariousness and uncertainty (see Podder and Zepeda Gil in Chapter 3).

Berlant is interested in the stories we tell ourselves and others in order to stay afloat. For Berlant, hope is structured and enacted through culturally specific genres of storytelling. Genres, they write, 'present an affective expectation of watching something unfold, whether that thing is in life or in art'(Berlant 2011, 6). While conventional genres of event help us to process events in their emergence, they can also 'foreclose the possibility of the event taking shape otherwise' (Berlant 2011, 6). Berlant thus identifies crisis as a particular genre of event:

> The genre of crisis is itself a heightening interpretive genre, rhetorically turning an ongoing condition into an intensified situation in which extensive threats to survival are said to dominate the reproduction of life. At the same time . . . the genre of crisis can distort something structural and ongoing within ordinariness into something that seems shocking and exceptional. (2011, 6)

Crisis, for Berlant, is a particular genre of event that marks a clear a distinction between normal and exceptional times – a distinction which Berlant ventures to blur in their own concept of 'crisis ordinariness'. Berlant's writings on 'crisis ordinariness' (2008) attempts to explain the desire amongst individuals to establish a sense of normality within a period of crisis. Crisis ordinariness refers to the enormous work of a subject to normalise crisis, rather than be traumatised by it. This may take the form of ignoring it, developing habits and routines to cope with its affects, and indeed to hope for a return to 'normality'. Overwhelmed by crisis, the subject retreats by instituting a form of 'affect management' that, whilst preserving the subject, serves to short-circuit, dispel, interrupt, moderate hopes of more fundamental changes. Hope persists, but it becomes 'cruel'.

Different genres of event may indeed play an influential role in promoting a desire for a return to normality, of 'affect management', in the wake of COVID-19. Take, for example, the genre of scandal as a particular mode of mediating our understanding of the experience of COVID-19. Within the United Kingdom, scandal was a dominant mode through which criticism of the COVID-19 response was articulated and presented to the public via the media. Jamie Johnson describes scandal as a normative mode of truth-telling deriving from a moral transgression (Johnson 2017). A scandalous act is one that is generally regarded as crossing a line from morally permissible to impermissible. The exposure of a scandal is an incitement to participate in the denouncement of what is portrayed as a clear moral violation of ethical norms that are both common (shared) and common-sensical (obvious). In this sense, the revelation of a scandal provides a restored sense of (moral) certainty in the midst of the confusion of a prolonged crisis. As a genre of event, the scandal can thus have considerable cathartic effect: establishing a sense of certainty, directing blame, focusing anger and exercising tension.

Within the circuits of UK news media, the genre of scandal appeared as a devastating form of critique deployed to hold power to account. 'Partygate', 'Beergate', 'Wallpapergate' and numerous other scandals (Halliday 2022; Siddique 2022) drew outrage from the public by aligning political failings in the UK's COVID-19 response with the moral failings of certain politicians. Yet, in foregrounding individual moral transgressions, the steady eruption of scandal deflects public attention from considering more complex questions surrounding the litany of failures which undermined the government's response. As a genre of event, scandals shape expectations by imposing a narrative on events within what Johnson calls a 'restrictive politics of recrimination' (2017, 722) that limits the possibilities of critical thought. In the case of the UK's COVID-19 response, the framing of failures through the moralising prism of scandals inhibited more nuanced discussions on the neoliberalisation of emergency response, the debilitating legacies of austerity in undermining resilience, and how corruption at the heart of procurement processes was facilitated by the extant logic of disaster capitalism which encouraged this crisis to be exploited as an opportunity for profit.

As a genre of event, scandal appears as a scathing critique. However, on closer inspection, the moral framing encourages a search for guilty individuals over considerations of the ethical problems at the heart of the systems comprising our contemporary order. The appearance of the scandal also has significant affective implications, abating the confusion associated with crisis by proving a familiar morality tale that restores a sense of ordinariness. At the same time, the scandal has a cathartic function allowing us, the public, to vent frustrations and anger at a guilty other (the corrupt politician), whilst absolving and distancing ourselves from any personal moral culpability. In doing so, scandal can provoke cynicism, reducing complex social, political and economic problems, which can be addressed politically, to a well-worn narrative of the inherent corruption of the political classes. It gives the impression of holding power to account, and provides a feeling associated with the realisation of justice that ultimately fails to challenge a failing social, political and economic order that has rendered lives so insecure. Scandal encourages the subject towards apathy, distraction and cynicism as a means of affective management, but it comes at the cost of envisioning and working towards the construction of a political system capable of delivering security and well-being.

## Conclusion: A Hope Against Hope

According to Koselleck, the conceptual coupling of crisis and critique was made possible by the emergence of a newfound understanding and experience of time, teleology and the event within Enlightenment thought. We might ask whether similar shifts are taking place presently. The advent of the Anthropocene has had a profound effect on how we understand and respond to crises. An appreciation of the non-linear becoming of such events discloses a world of contingency, unpredictability, and the entangled multiplicity of polycrises. Faced with the incapacity to predict and prevent crises, the response has been to build our resilience to them (Chandler 2014; Zebrowski 2016). Yet this is insufficient. The 2014 Intergovernmental Panel on Climate Change report warns:

> Without additional mitigation efforts beyond those in place today, and even with adaptation, warming by the end of the 21st century will lead to high to very high risk of severe, widespread, and irreversible impacts globally. (2015, 17)

Even if carbon dioxide ($CO_2$) emissions were completely halted today, the effects of current emissions are predicted to yield 2.7 degrees Fahrenheit of warming which are more or less inevitable (World Bank 2014, xvii). The advent of the Anthropocene demands much more urgent and ambitious action to address the intensifying problems of climate change which are swiftly becoming existential (Kumar 2023). Bruno Latour (2020) has suggested that the international response to COVID-19 was a 'dress rehearsal for climate change'. It was, and the conclusions are not great. Rather than take stock of the lessons learned from the pandemic and 'build back better' to prepare for the devastating polycrises anticipated in the future, the general thrust of the response has simply been to restore a pre-COVID sense of 'normality'.

This chapter has sought to explore why, in light of the profound vulnerabilities, injustices and insecurity exposed by the COVID-19 polycrisis, the deep abiding hope at present for many is simply for a return to normality. Drawing on Berlant's concept of 'crisis ordinariness', I have been compelled to try and understand the desire for normality as a form of cruel optimism: a mode of 'affective management' which helps to abate the confusion and anxiety experienced by the subject in a time of crisis, but which is ultimately incapable of (or even counterproductive for) delivering the conditions necessary for the realisation of the 'good life'. In this vein, I argued that the British tendency to frame critical responses through the moralising genre of scandal, while providing a semblance of familiarity and redress to the British public, in fact delimited and diluted the critique of the United Kingdom's COVID-19 response by deflecting attention from more structural factors, including legacies of inequality, racism and austerity, that rendered the British public (and certain populations vastly more than others) vulnerable to these kinds of events. Scandal, as such, serves to bolster a cynical disposition by locating the root of our problems in the intractable moral failures of politicians.

Truthfully, the conditions for hope in the Anthropocene appear depleted. We are told that the chances of returning to a familiar 'normality' are likely already gone. Instead, we must prepare for a world that is likely to be very different than the one of recent memory. But even if the Anthropocene dashes modernist hopes that advances in knowledge might lead to human mastery of the environment and achieve stability and security, it should not, and indeed cannot, extinguish hope entirely. Hope persists, even if that requires it to turn cruel: to submit to fantasy, denialism, apathy and cynicism. If the critical project of modernity was to establish the grounds and limits of reason, perhaps what is required most critically today is to cultivate, revive and reorient the conditions for hope within the Anthropocene: to hold out a hope against hope.

The English idiom 'a hope against hope' means to sustain hope in a hopeless situation. It means retaining hope in the face of almost certain disappointment, often without reason or justification. To hold out a hope against hope in the Anthropocene might mean finding ways of cultivating, sustaining, reinvigorating and reorienting hope in the face of an increasingly uncertain and insecure future. This may well mean fostering a hope for something beyond a return to normality. Such a project would no doubt be dependent upon a radical redefinition of ideas of hope (too idealistic) and critique (not sufficiently creative) from the ways in which these terms figured within Enlightenment narratives. Safeguarding hope from descending into cynicism would require a relentless critique of utopian projects promising salvation. But it might also, following Berlant (2011), call for the formulation of new genres of storytelling that do not just provide fictional sustenance by way of distraction and catharsis, but that could also cultivate the conditions under which we could identify new ways to narrate our lives that sustained better ways of living. To hold out a hope against hope is to recognise the dogged persistence of hope even as it manifests within conditions of diminishing returns, and to explore new narratives which reinvigorate thought, life and political action in uncertain and insecure times.

What can be hoped for? Hope itself.

## Notes

1. For prominent scholarly debates over when to date the beginning of the Anthropocene, see Lewis and Maslin (2015).
2. Thanks to Valerie Waldow for this point.

## References

Anderson, B., Grove, K., Rickards, L. and Kearnes, M. 2020. Slow emergencies: Temporality and the racialized biopolitics of emergency governance. *Progress in Human Geography* 44(4): 621–39.

Berlant, L. 2008. Thinking about feeling historical. *Emotion, Space and Society* 1(1): 4–9.

Berlant, L. 2011. *Cruel Optimism*. Durham, NC: Duke University Press.

Bloch, E. 1996. *The Principle of Hope*, vol. 1. Cambridge, MA: MIT Press.

Briggs, D., Ellis, A., Lloyd, A. and Telford, L. 2020. New hope or old futures in disguise? Neoliberalism, the Covid-19 pandemic and the possibility for social change. *International Journal of Sociology and Social Policy* 40(9–10): 831–48.

Chakrabarty, D. 2018. Anthropocene time. *History and Theory* 57(1): 5–32.

Chandler, D. 2014. *Resilience: The Governance of Complexity*. London: Routledge.

Chandler, D., Grove, K. and Wakefield, S. (eds). 2020. *Resilience in the Anthropocene: Governance and Politics at the End of the World*. Abingdon and New York: Routledge.

Cole, J. and Dodds, K. 2021. Unhealthy geopolitics: Can the response to COVID-19 reform climate change policy? *Bulletin of the World Health Organization* 99(2): 148–54.

Colebrook, C. 2014. *Death of the PostHuman: Essays on Extinction*, vol. 1. London: Open Humanities Press.

de Man, P. 1967. The crisis of contemporary criticism. *Arion: A Journal of Humanities and the Classics* 6(1): 38–57.

Drezner, D. 2023. Are we headed toward a 'polycrisis'? The buzzword of the moment, explained. *Vox*, 28 January. Available at https://www.vox.com/23572710/polycrisis-davos-history-climate-russia-ukraine-inflation (accessed 21 December 2023).

Gibson-Graham, J. K. 1996. *The End of Capitalism (As We Knew It): A Feminist Critique of Political Economy*. Minneapolis: University of Minnesota Press.

Gross, J. 2022. Hope against hope: COVID-19 and the space for political imagination. *European Journal of Cultural Studies* 25(2): 448–57.

Halliday, J. 2022. Scandal after scandal: Timeline of Tory sleaze under Boris Johnson. *The Guardian*, 1 July. Available at https://www.theguardian.com/politics/2022/jul/01/scandal-timeline-tory-sleaze-boris-johnson (accessed 21 December 2023).

Haraway, D. 2016. *Staying with the Trouble: Making Kin in the Chthulucene*. Durham, NC: Duke University Press.

Intergovernmental Panel on Climate Change. 2015. *Climate Change 2014 Synthesis Report*. Geneva.

Johnson, J. M. 2017. Beyond a politics of recrimination: Scandal, ethics and the rehabilitation of violence. *European Journal of International Relations* 23(3): 703–26.

Juncker, J.-C. 2016. Speech by President Jean-Claude Juncker at the Annual General Meeting of the Hellenic Federation of Enterprises (SEV) [Text]. European Commission. Available at https://ec.europa.eu/commission/presscorner/detail/en/SPEECH_16_2293 (accessed 21 December 2023).

Kolbert, E. 2014. *The Sixth Extinction: An Unnatural History*. New York: Henry Holt.

Koselleck, R. 2006. Crisis. *Journal of the History of Ideas* 67(2): 357–400.

Kumar, A. 2023. Ruptures of the Anthropocene: A crisis of justice. *Dialogues in Human Geography* 13(2): 202–6. https://doi.org/10.1177/20438206231155704.

Latour, B. 2020. Is this a dress rehearsal? *Critical Inquiry*. Available at https://critinq.wordpress.com/2020/03/26/is-this-a-dress-rehearsal/ (accessed 21 December 2023).

Latour, B. and Porter, C. 2017. *Facing Gaia: Eight Lectures on the New Climatic Regime*. Cambridge and Medford, MA: Polity Press.
Lewis, S. L. and Maslin, M. A. 2015. Defining the Anthropocene. *Nature* 519(7542): 171–80.
Marcel, G. 2007. *Being and Having*. [no place]. [no publisher]. Available at http://dhspriory.org/kenny/PhilTexts/Marcel/BeingAndHaving.pdf (accessed 21 December 2023).
Morin, E. and Kern, A. B. 1999. *Homeland Earth: A Manifesto for the New Millennium*. Cresskill, NJ: Hampton Press.
Nixon, R. 2011. *Slow Violence and the Environmentalism of the Poor*. Cambridge, MA: Harvard University Press.
Roitman, J. 2014. *Anti-Crisis*. Durham, NC: Duke University Press.
Serhan, Y. 2023. Why 'Polycrisis' was the buzzword of day 1 in Davos. *Time*, 17 January. Available at https://time.com/6247799/polycrisis-in-davos-wef-2023/ (accessed 11 March 2023).
Siddique, H. 2022. Scandalous legacy: As Johnson heads for the exit, many issues remain unresolved. *The Guardian*, 9 July. Available at https://www.theguardian.com/politics/2022/jul/09/scandalous-legacy-as-johnson-heads-for-the-exit-many-issues-remain-unresolved (accessed 21 December 2023).
Sloterdijk, P. 1987. *Critique of Cynical Reason*. Minneapolis and London: University of Minnesota Press.
Stanley, S. 2007. Retreat from politics: The cynic in modern times. *Polity* 39(3): 384–407.
Šumonja, M. 2020. Neoliberalism is not dead – On political implications of Covid-19. *Capital and Class*, 28 December.
Tooze, A. 2021. *Shutdown: How COVID Shook the World's Economy*. New York: Viking.
Tooze, A. 2022. Welcome to the world of the polycrisis. *Financial Times*, 28 October. Available at https://www.ft.com/content/498398e7-11b1-494b-9cd3-6d669dc3de33 (accessed 21 December 2023).
Tooze, A. 2023. Three ways to read the 'deglobalisation' debate. *Financial Times*, 30 January. Available at https://www.ft.com/content/b3f41263-88d9-4012-aafc-145f0327678f?utm_medium=email&utm_source=substack (accessed 21 December 2023).
Walby, S. 2021. The COVID pandemic and social theory: Social democracy and public health in the crisis. *European Journal of Social Theory* 24(1): 22–43.
World Bank. 2014. *Turn Down the Heat*. Washington, DC: World Bank Group.
Zebrowski, C. 2016. *The Value of Resilience: Securing Life in the 21st Century*. London: Routledge.
Žižek, S. 2020. *Pandemic! Covid-19 Shakes the World*. New York: OR books.

# 3

# Working for 'Minor Utopias': Youth Employment in Sierra Leone and Liberia

*Sukanya Podder and Raúl Zepeda Gil*

### Imagining Youth Futures in Post-war Societies

For the young, hope is an exercise in active imagining. It is co-produced, on the one hand, by how they perceive the structural, socio-economic opportunities or lack thereof and, on the other hand, by how the actors, institutions or organisations may support or inhibit the futures they imagine. Post-war is a crucial moment for fresh hope amidst the problems of transition to stability and a secure political settlement. Whilst war presents pathways to violent political change, it encourages youth[1] involvement as soldiers and civilians (Hutchinson and Inayatullah 2010). War can constrain youth's capacity and agency to choose alternative destinies amidst war-induced displacement and upheaval.

The end of wars opens up new possibilities about post-war futures, which are co-produced through the post-war development and pacification processes. The end of a conflict opens several streams of hope and imagination for the youth – although some of these idealisations depend on the character of the peace settlements, or even, as Pospisil (2019) argues, (un)settlements that give the impression of peace even when some legacies of conflict remain. Critical moments of political breakthrough open the possibility to view new forms of employment in a restored economy as a way to more secure and stable livelihoods. In particular, for youth in a transition phase into adulthood, rather than an age-based perspective on how life occurs, there are critical junctures for their transition into independent adult life (Elder 1998).

These junctures include detaching from the household, forming families and leaving school. War and the foreclosure of a 'regular' future hinder these prospects. In the post-war period, economic crises or other socio-economic inequalities continue to shape youth transitions into gainful employment. During the war years, joining rebel groups as soldiers became a form of wartime labour rather than an abnormal upending of childhood (Hoffman 2011). Similarly, during the post-war period, most youth may choose work over education and a return to schooling. These choices express the need for survival and present a less graduated transition to adulthood. In this sense, war and post-war societies affect the usually expected pathways to adulthood for young people followed in peaceful Western liberal societies (Cooper et al. 2019).

Post-war hopes are shaped by youth decisions in the context of economic hardships and structural constraints. Crucially, employment plays a role in young people's minds, and their decisions are premised on their expectations about salary, location, sustainability and the possibility of progression. As Breen and Goldthorpe (1997) explain, based on each person's information, employment and education-related choices are constructed on the imagined possibilities of upward social mobility, belief in their capacities, and expectation of possible future outcomes. Particularly in the Global South, these choices are also intertwined with concerns around precarity, informality, caring demands and limited social services (MacDonald 2009).

Beyond the classical rational choice perspective, three caveats are necessary for this decision-making process towards employment and education for youth. First, these decisions are bounded by insufficient information, constraints and, most importantly, by how each individual perceives these factors (Elster 1976; Simon 1990). Second, the centrality of existing choices for youth depends on how they can exercise their agency beyond parental or governmental oversight. However, as seen in youth studies scholarship, this agency is bounded by the conditions of possibility (Evans 2002). Third, in the post-war context, navigation of available opportunity structures depends strongly on socio-material conditions. Therefore, youth decisions follow a process called structuration, in which agency and existing structures shape each other (Giddens 1986).

Instead of interpreting agency as a deterministic product of conditions and structures, this chapter analyses everyday choices towards social mobility as a form of negotiation with the structures in which different types of hope play into the imaginaries of youth. In the eyes of young people, hope is how these interactions are dealt with. As Cahill (2015) mentions, youth voices express how the future is constructed, allowing them to be the experts in their lives. Notably, this means embracing the diversity of life courses as perspectives rather than as fixed endpoints. Although employment, reproduction, housing and education are markers of social mobility, these do not manifest in a linear sequence but through a messy and complicated process of everyday muddling (Furlong 2015). The only way to grasp the diversity of hope(s) linked to multiple imagined futures is to accept their plurality and their often contradictory nature and to observe how these categories of hope intersect or diverge.

**Agency and Hope**

David Chandler's analysis of hope in Chapter 1 offers some essential nuances to structuralist analyses while opening space for agency and negotiation. Chandler introduces three categories of hope – *speculative*, *pragmatic* and *nihilist* – that apply to youth futures in post-war contexts in the following ways. First, there is the youth's speculative hope. This category emerges from the endless possibilities of a new kind of society beyond the constraints of civil conflict and colonialism. In particular, unravelling conflict brings the possibility of a new post-war generation. This has the potential for longer lifespans uninterrupted by war and also the chance to mature in a socio-political milieu devoid of the triggers that led to the previous conflict. In this sense, speculative hope is transformative. It can be expressed in the fertile ground of nothingness but, most importantly, in the context of reduced or no violence, which presents the opportunity for peaceful lives and livelihoods. Effectively, post-conflict societies do not become blank slates of perfect harmony but remain in an evolving state of political (un)settlement. Regarding social mobility and employability, it simply means having the space to recreate professions or imagine ones never thought of before.

At the same time, youth's hope for the future must be pragmatic. Conflict legacies and peace settlements constrain their speculative hope, even if limited violence and new structural realities are being negotiated (Leonardsson and Rudd 2018; Mac Ginty and Richmond 2013). The new period of post-war peace offers at least the potential for emancipatory agency and the ability for youth to renegotiate opportunity structures. This means that youth have the potential to inform and shape the existing aid programmes, development policies and peace initiatives through their own hands, transforming them into their aspired futures. Various peacebuilding and international development programmes targeting youth offer new training and learning opportunities. Employment programmes offer pragmatic choices regarding livelihood generation and encourage entrepreneurship, allowing youth to build their capacities and work hard to achieve upward social mobility (Peeters 2009).

Finally, nihilist hope can also play into youth imaginaries to endure the bleakness of the post-war future. Even though peace is formally reached, nihilist realism lies in accepting the world as it is by embracing the precarity of the present and the near foreseeable future. Many post-war countries do not thrive economically, making everyday survival and pessimism part of youth futures. From a social mobility prospect, post-conflict countries are often the location of conditions for new forms of precarious employment, lack of opportunity and chronic underemployment (Fennell 2020; MacDonald and King 2021). Peace without development reproduces the structural conditions often at the roots of conflict, making it necessary to redress the inequalities that can remain unresolved even decades after one episode of conflict ends (Galtung 1969).

These forms of hope are dependent on how the political economy of a post-war country interacts with the agency that youth can exercise, within structural boundaries, to seek social mobility and how their imagined futures adapt to different life outcomes – from thriving to resilience or even disappointment. Nihilistic youth futures have been historically related to the marginalisation or the lack of spill-over effects of post-war aid and development into tangible trickle-down effects in young people's everyday lives. In opposition, another hope is regained through a collaborative approach: a hands-on, youth-led and adult-supported process (Podder et al. 2021). This type of hope moves expectations from short-term and short-sighted goals, usually informed by moral panics around youth as a source of risk, to more long-term and future-oriented programming. Mac Ginty (2021) shows that even in the most desperate war scenarios, small-scale community actions can interrupt violence and create alternative visions of the present and the future.

**Minor Utopias and Youth Futures**

In the context of stagnant post-war economies, social mobility choices are limited and hopeful futures concerning employment can constitute what Winter (2006) calls 'minor utopias' – 'imaginings of liberation on a smaller scale, without the grandiose pretensions or the almost unimaginable hubris and cruelties of the "major" utopian projects' (Winter 2006, 5). Therefore, rather than encompassing world transformations through ideological projects such as socialism, fascism or global liberalism, the two cases we share in this chapter arise as small-scale imaginaries from below of better futures in the context of the fundamental uncertainty and despair that can constrain youth choices. We propose that youth engage in an intellectual task that is also a collaborative, felt experience. Youth create these futures from their lived experiences. In these processes, future dreaming or visualisation efforts require engaging with potential solutions by reframing hopelessness into action and giving youth agency and capacity for seeking real solutions. The categories of hope examined earlier are usually enacted as minor utopias when these hopes are not realised due to structural limitations. As will be shown later, in the cases of

Liberia and Sierra Leone, there are several lessons about youth futures, future dreaming and the structural barriers that translate hopes into minor utopias.

Hutchinson and Inayatullah (2010) have mapped the future literature and its intersection with peace studies. They demonstrate important types of futures for analysis. When the new scenario is open, three dominant views of the future (realist, consumerist or authoritarian) are imposed or infused; they offer nihilistic or journey views, which are open to many scenarios ranging from pessimism to optimism to more critical views of the future, either politically dissident or transformatively pluralistic. From these views, the futures of social mobility and forms of hope interlink several scenarios where youth can utilise the apparent 'blank slate' of the post-war reality. From a future-dreaming perspective (Cahill 2015), the numerous intersections and possibilities depend on adult-supported processes, graduated programming and long-term commitment to youth development.

Stewart (2009) has found in several post-war countries that economic inequalities remain untouched after peace settlements and even after implementing development programmes. As Chandler (2006) argues, peace agreements and post-war state-building inherit old elites and create dependency on Western governments' development aid. The top-down approach to peacebuilding fosters scenarios in which the heavy burden of the international and local hierarchies can impose either hopelessness or nihilistic hope. Therefore, Inayatullah (2002) sees the creation of youth futures as disrupting and challenging these hierarchies and conditions by not conforming to structures, muddling through them, and even posing a rebellious position. This repertoire of the future dreaming approach can transcend individualistic approaches to technology (social media), protest and self-organisation into civic movements.

A relevant feature of the youth futures paradigm is the denial of the adult-centrism built into the structures of the status of civilisation. As Bussey (2002) argues, the production of youth futures challenges the dominant civilisational portrayals, not only from the perspective of a world that the current youth generation will inherit but also by establishing the complex intergenerational debate that constantly happens in cycles of human history. Repeatedly, the repertoire above of future-building scenarios for social mobility also implies defying views of the world of work. For example, the diffusion of ideas via the World Wide Web and horizontal social media networks has increased to unprecedented levels. In some scenarios, these exchanges have also created the scope for intergenerational dialogue on issues such as climate change and the future of work (Djohar and Pruitt 2021).

Elaborating on the role of intergenerational exchanges for intergenerational futures, Johnston-Goodstar (2020) suggests that the role of youth in United

States Indigenous communities has come to be re-evaluated by challenging curriculum design and community values around childcare. This revision offers youth a future that can be envisioned in a decolonial perspective, not dependent on external actors' views, but rather in the context of their own cultures and by adapting to their everyday lives (see also Nicoson in Chapter 5).

Alternatively, in India, Koskimaki (2017) argues that in the Uttarakhand region young men have transformed the views of their communities about the future of youth work and dependency on adults for more positive futures by using print media and rallies to describe their region as a place where they can aspire to social mobility.

One key element underlying this conceptual discussion around hope and youth futures is uncertainty. The building of the future implies facing uncertainty about the results. Due to the inequalities dominating the world, the axiomatic approach to justice implies that it is possible to search for and be hopeful about opportunities and future life paths without constraints (Roemer 1998; Sen 1999). However, deep structural and socio-economic inequalities around wealth, education, race and nationality bring different levels of uncertainty. Therefore, building youth futures in post-war environments must navigate these layered realities. As Black and Walsh (2019) found with university students in Western societies, the futures that they built unfolded in three ways: their individual futures (microfutures), the immediate futures in their countries (mesofutures) and the perceived global changes (macrofutures) or zeitgeist. In Africa and other parts of the postcolonial world, however, conflict creates further challenges, and youth's hopeful visions can become reduced to minor utopias when they deal with the realities of work and hope in post-conflict societies.

**Youth, Work and Hope in West Africa**

For decades, Africa's youthful population could expect three outcomes as they prepared to enter the labour market: first, success in finding secure formal sector employment through a personal network or family contacts, patronage being more important than merit; second, for those with limited contacts in government, contentment with less secure formal sector employment in the form of part-time or rolling contracts or through entrepreneurship, especially in technology-driven sectors (Sumberg et al. 2019, 429–31); third, for the marginalised and poor, resignation to 'getting by' through hustling for daily wages in the informal economy (Podder 2015). These pathways aptly capture the inequality of opportunities and the differences and the types of hopes youth could access for upward social mobility.

The last decade has seen an explosion of policy and public interest regarding youth employment across sub-Saharan Africa. This increased attention can be linked to heightened awareness of the continent's 'youth bulge' (Baah-Boateng 2016) and the interest in securing an associated 'demographic dividend'. Following a decade of sustained economic growth that failed to create significant numbers of new formal sector jobs, the possibility of youth unemployment leading post-conflict countries like Liberia and Sierra Leone back to violence and political instability becomes more prominent (Brück et al. 2016). In response, policy and development interventions have focused mainly on the young people themselves, particularly as they move into the workforce. In addition to investment in general education, technical and vocational education and training remain a common intervention area, along with employability (soft) skills and entrepreneurship training. The notion of a 'skills gap' underpins these interventions (World Bank 2017). Early analyses of these policies viewed youth from the lens of adults' expectations (Sommers 2011).

Fortunately, new approaches in development policy have shifted towards youth agency and new co-produced futures. In the following case study analyses, we focus on how far a shift away from interventions that focus on the supply side and towards the demand side of the labour market has created a hopeful position towards achieving 'decent work' for young people that is central to the Sustainable Development Goals (Sumberg et al. 2019, 430). We examine the 'Youth Employment and Empowerment Programme' in Sierra Leone and the different 'Youth Employment Programmes' in Liberia, developed and funded by national and international agencies like the United Nations Development Programme (UNDP) in partnership with international and local civil society partners.

## Sierra Leone and Youth Empowerment

According to one estimate, 70 per cent of Sierra Leone's youth are unemployed or underemployed. Youth, defined in Sierra Leone as people aged 15 to 35, account for 34 per cent of the national population, estimated at six million. *Under*employment or poor working conditions and low pay is more about the quality of the job rather than the absence of a job (Page 2013). A decent job and the economic independence and personal empowerment that come with it are often critical markers in the transition to adulthood. In Sierra Leone, the 1991–2002 civil war resulted in changes in its economy. It is well known that wars can accelerate processes of economic transformation through restructured access to land and labour markets (Wood 2008). Civil wars also cause large-scale forced displacement of civilian populations, and many suffer death, injury, loss of land and property, and disrupted schooling, thereby affecting the demographic characteristics at a local level. In post-war societies, peacebuilding involves physical reconstruction, social reconciliation and the transformation of war economies into peace economies. In this sense, one of the goals is to change the incentive structures from engaging in violent 'labour' as combatants to peaceful labour for ex-combatant and non-combatant populations (Ballentine and Nitschke 2005; Hoffman 2011).

In Sierra Leone, youth-focused peacebuilding is rooted in an overconfidence in employment creation as the panacea for peaceful intergenerational relations. Successive governments have placed youth employment and empowerment at the heart of their peacebuilding discourse, yet, eighteen years after the war ended, approximately 70 per cent of youth remain unemployed or underemployed, with widespread illiteracy, educational opportunities being beyond the reach of many. Employment generation and livelihood support has ranged from short-term projects focused on vocational training to graduate internship schemes sponsored by the UNDP under the Youth Employment and Empowerment Programme, a US$9 million donor-funded effort. It involves a Graduate Internship Programme initiative of the National Youth Commission of Sierra Leone supported by the UNDP. The National Youth Commission (NYC) has identified a lack of work experience as a significant obstacle for job-seeking graduates (Alemu 2017; UNDP Sierra Leone 2016, 3–5). With UNDP's technical and financial assistance, the NYC initiated a structured pilot graduate internship programme that targeted qualified graduates wishing to acquire formative and productive work experience. Restless Development, an international non-governmental organisation (NGO), has developed a professional internship model and placed interns across private, public and non-governmental institutions throughout Sierra Leone (UNDP Sierra Leone 2016).

Since 2012, seventy-one governmental institutions, civil society organisations and private businesses in Sierra Leone have partnered with UNDP and NYC in this programme. The Graduate Internship Programme involves a three-month internship placement in various organisations that allows youth to gain work experience and a stipend of US$100 per month. Five Career Advisory and Placement Services (CAPS) centres have been established in various locations, such as the Njala and University of Sierra Leone campuses. These centres aim to provide university and college students with labour market and career information, job search skills, IT training, job placement opportunities and career workshops. They have also promoted constructive relationships between educational institutions and potential employers. A forum for fifty employers was organised to allow them to make direct input on demand for specific skills. According to 2018 figures, 4,625 (2,594 male and 2,031 female) university students benefited from services available at the various CAPS centres (Alemu 2017).

The National Youth Commission, with the support of the UNDP, has also developed Business Development Service Centres that opened in 2012 in five main urban centres: Freetown, Newton, Bo, Makeni and Kenema. Each centre offers direct assistance to over 2,000 entrepreneurs per year. The aim has been to create and expand sustainable enterprises, leading to increased employment opportunities for young people – 983 employed (597 male, 386 female) due to interventions by the business development services. According to 2018 figures, 216 (120 male and 96 female) business enterprises established by youths have been registered with government authorities, improving the legal environment in business start-up processes; in addition, 3,298 (1,734 male and 1,564 female) business operators have benefited from business development services' mentorship, advisory, coaching and business outreach education information. Despite the best intentions, the scale of these efforts is simply not enough. In 2016, over 1,000 applications were received for four hundred placement positions in the insurance scheme alone (UNDP Sierra Leone 2011).

Similarly, the Youth Employment Support (YES) project funded by the World Bank was designed with an eye to regional sensitivity to ensure that a specific quota of young people in every region of Sierra Leone could be enrolled in it. The programme also encouraged female participation, to the tune of 40 per cent out of a total beneficiary sample of 9,000 youths. To be more representative of youth diversity, the YES programme targeted three categories of youth: 'youth with low levels of education', 'youth with high levels of education and 'youth within rural areas'. Although the project was well thought through, the simple fact was that the target of 9,000 youth was minimal, given the large numbers of youth requiring similar support (Podder et al. 2021, 3). Like YES, most peacebuilding and development projects focused on youth employment have a limited scope; they are designed for a specific target population or a sample thereof. Therefore, even if programmes like the YES project were to continue for a long time, they could only help a few young people. The results of these approaches demonstrate the problem of how limited growth and a challenging political economy either restrict hopes of social mobility or produce a muddling through of nihilistic futures.

## Liberia and Youth Employment Programming

Similar safety net projects in post-war Liberia such as the Youth, Employment and Skills Project, the Liberia Youth Employment Program, and, more recently, the Youths Employment Project have offered stop-gap employment support for a small percentage of the youth population. For example, the Youths Employment Project provided three months of employment to 2,500 young people between 18 and 35 years of age. Each beneficiary received US$60 per month for unskilled work, while US$100 per month is paid for skilled labour. The Youth Employment Project was implemented in Grand Bassa, Margibi, Bomi, Grand Cape Mount and Montserrado. The Liberian Agency implemented the intervention for Community Empowerment under the overall institutional oversight of the Ministry of Labour and the Ministry of Public Works (ILO 2014, 7–9).

The Youth, Employment and Skills or YES project, which ran between 2010 and 2013, had two main components. The first was to create temporary employment via community-based public works. This pillar had two sub-components: targeting 45,000 Liberians to offer temporary (40 days) employment and targeting 'at risk' youth. The second sub-component was to build the Government of Liberia's capacity to undertake and coordinate similar activities. The second YES component funded skills training programmes to improve youth employability, including providing institutional support to technical and vocational training. The Liberia Youth Employment Program was introduced in February 2013. Like the YES, it aimed to employ vulnerable youth (18–35 years) and sought to increase employability. It expanded employment for an entire year and was implemented by the Ministry of Youth and Sports until 2016 (ILO 2014, 7–9).

Despite these efforts, there are still limited income-generating opportunities. According to 2019 figures, only 18 per cent of the workforce was paid, and the informal sector accounted for 85 per cent of all employment in the country. Relating to the labour market characteristics, the labour force participation rate for the 15–24 youth cohort in 2012 was only 35 per cent, lower than the national average of 63 per cent. To boost these figures in 2013, the government allocated US$75 million over five years to address youth unemployment (i.e. US$15 million per fiscal year). This funding, in addition to international assistance, signalled a solid political and financial commitment to the matter (ILO 2014). Scholars have shown that providing jobs and youth employment was paramount in stabilising post-war Liberia (Munive 2010, 323). The overall problem lies in how the structural changes in the economy could provide diverse choices to empower the speculative hopes of the youth in the region.

A decade later, in 2023, the latest UNDP figures suggest that thus far, and many millions of dollars of investment later, only two sectors of the economy provide sustainable income and livelihoods, namely agriculture and informal micro, small and medium enterprises (MSMEs). These sectors are facing several challenges. The agriculture sector is constrained by low human capital, poor infrastructure (roads, access to energy, machinery and technology), poor access to farm inputs, credit and extension services, weak policies that deter private investments, and a lack of trust among market actors to engage in bulk purchasing, storage and marketing. On the other hand, MSMEs are constrained by weak management, fragmented markets, limited diversification, high operating costs, limited access to capital and credit, a limited skilled workforce, and the absence of long-term planning, among others (UNDP Liberia 2023).

Donors and agencies like the UNDP are shifting their rhetoric from offering short-term jobs to gradually transforming employment opportunities into medium-term jobs, long-term employment and inclusive growth. However, this remains aspirational. It would require concerted planning, structured investments in developing an agro-inputs supply chain and providing technical advice to increase local food production (UNDP Liberia 2023). It also requires rethinking access to land and its use for youthful populations (Podder 2023). Achieving these goals will take time and long-term commitment from the national government and international investors. Despite this shift in rhetoric, a fundamental limitation plagues these endeavours. This pertains to the limited appreciation of futures-oriented thinking in employment support, rooted in intergenerational responsibility. Usually, donors, NGOs and national governments offer youth ad hoc access to mentorship, materials and technical support rather than more sustainable long-term employment policies. Notably, this economic transformation has yet to produce sustainable formal employment that can foster social mobility beyond precarious and seasonal challenges that the youth frequently face.

### *Bike Riding and Minor Utopias in West Africa*

Both the Sierra Leone and Liberia cases show the internal problems and complications from top-down approaches to employment aid in peacebuilding for African youth. In these contexts of precarity, new minor utopias have surged – for example, in motorcycle taxis or bike riding as a way of entrepreneurship and livelihood. Jenkins et al. (2020) have found an increasing female interest in becoming motorcycle taxi drivers, the most important way of transport in rural areas of Sierra Leone and Liberia. These motorbikes have restored mobility in rural, landlocked areas of both countries. International aid focused on infrastructure building in areas where motorcycle taxis ply has also gained traction. Since early 2017 the German Development Agency funded the Gogein track network construction, and in the process, young women were actively involved in the construction and became one of the leading groups interested in operating bikes, or *okada*s as they are locally called. Although it is not without risk of harassment, female drivers have brought new, unimagined employment for young women seeking work opportunities outside their residential areas. In this case, we see the construction of minor utopias through pragmatic hopes that take advantage of foreign aid.

Also, as McMullin (2022) has found, several groups of demobilised former youth combatants have integrated themselves as motorcycle drivers in the two countries after the civil war. An account of speculative hope can be gleaned from the 'Best Man Corner Station' located northeast of Monrovia, Liberia's capital city. These young men consider their occupation a proud endeavour that brings identity and purpose. Their everyday accounts show how they can envision futures beyond precarity by working hard instead of engaging in robbery or crime like some demobilised soldiers. Other youth demobilise and now, in motorcycle driving, have constructed nihilistic minor utopias: it is just a job to eat and survive each day. For example, the young boys of Jallah Town do not seem to have a clear identity or pride in their chosen occupation; they express relief that they are not surviving the hail of bullets on the battlefield. Overall, these boys resent the criticisms towards them. Adults often criticise their occupation of bike riding as spoiling their future; however, some see motorcycling as a transitory occupation towards a more respectable future. For some, a motorcycle is an identity because it was their first occupation after the end of the civil war.

## Conclusion

Meaningful youth engagement in post-war societies demands a long-term commitment to youth-led and adult-supported processes that emphasise youth inclusion in the formal sphere and not simply donor-facilitated participation in short-term projects. It also requires adopting an intergenerational peace lens (Podder 2022). Using insights from recent theorisation around hope, nihilism and youth futures, we analysed the empirical material from the case of postwar Liberia and Sierra Leone. The empirical findings challenge development intervention in post-conflict societies to create alternative futures for youth. Initial expectations of the possibilities of work after conflict and the arrival of employment programmes as part of long-term peacebuilding strategies that could open alternative futures for youth have fallen short. Initial speculative hopes of youth change have been impacted by failure of large-scale employment generation policies that offer short-term training and internship opportunities to only a minuscule portion of the youth entering the labour market. Whether these interventions bring nihilistic or pragmatic hopes depends on the question of access to the scarce opportunities for employment that national governments, NGOs and international organisations offer.

In the case of Liberia, the uplifting of pragmatic hopes is followed by disillusionment for those trained and schooled, because meaningful employment or decent jobs are available to few applicants. Furthermore, in Sierra Leone, marginalised youth are in temporary employment training. Both schemes are based on the demand-side labour policies of training without pairing with the structural conditions to create permanent and progression-oriented career-based employment. Instead, youth have embraced precarity and engaged in speculative hope through bike riding. In other words, the macrofutures offered after the peace processes in both countries from top-down interventions around employment generation have not grappled with the structural conditions that constrain the choices, the nature of adult support and the needs of the current and future generations. Without these considerations, it is hardly possible for post-war governments to create new youth futures that can shift the narrative from an unreachable to a hopeful peace. However, it creates the space for entrepreneurship and adaptive processes of engaging with work that is not merely a minor utopia but one that presents a hopeful future.

## Note

1. Definitions of who is a young person, or a youth, vary in different social and political contexts. The UN General Assembly (UNGA) has defined 'youth' as the age between 15 and 24 years. However, there is no single agreed definition. For example, the lowest age range for youth is 12 years in Jordan, and the upper range is 35 years in some African countries. The World Health Organization (WHO) and UNICEF use the term 'adolescent' for those aged 10–19 years and 'young people' for those aged 20–24 years.

## References

Alemu, M. M. 2017. Career advice and placement services in Sierra Leone. International Policy Centre for Inclusive Growth. Available at http://www.ipc-undp.org/pub/eng/OP356_Career_advice_and_placement_services_in_Sierra_Leone.pdf (accessed 27 December 2023).

Baah-Boateng, W. 2016. The youth unemployment challenge in Africa: What are the drivers? *The Economic and Labour Relations Review* 27(4): 413–31.

Ballentine, K. and Nitschke, H. 2005. *Beyond Greed and Grievance: Policy Lesson from Studies in the Political Economy of Armed Conflict.* New York: International Peace Academy.

Black, R. and Walsh, L. 2019. *Imagining Youth Futures*. Singapore: Springer.

Breen, R. and Goldthorpe, J. H. 1997. Explaining educational differentials: Towards a formal rational action theory. *Rationality and Society* 9(3): 275–305.

Brück, T., Ferguson, N.T., Izzi, V. and Stojetz, W. 2016. *Jobs aid peace: A Review of the Theory and Practice of the Impact of Employment Programmes on Peace in Fragile and Conflict-Affected Countries*. Report for ILO, PBSO, UNDP and WBG, ISDC. Berlin.

Bussey, M. P. 2002. From youth futures to futures for all: Reclaiming the human story. In J. Gidley and S. Inayatullah (eds), *Youth Futures: Comparative Research and Transformative Visions*. Santa Barbara, CA: Greenwood Publishing Group, 65–78.

Cahill, J. 2015. Youth futures: Enabling youth voices – the how and the why. *World Future Review* 6(4): 485–8.

Chandler, D. 2006. *Empire in Denial: The Politics of State-building*. London: Pluto Press.

Cooper, A., Swartz, S. and Mahali, A. 2019. Disentangled, decentred and democratised: Youth studies for the global South. *Journal of Youth Studies* 22(1): 29–45.

Djohar, Z. and Pruitt, L. 2021. Intergenerational feminist peace: Global research and a case study from Aceh, Indonesia. In T. Väyrynen, S. Parashar, È. Féron and C. C. Confortini (eds), *Routledge Handbook of Feminist Peace Research*. London: Routledge, 333–42.

Elder Jr, G. H. 1998. The life course as developmental theory. *Child Development* 69(1): 1–12.

Elster, J. 1976. *Ulysses Unbound: Studies in Rationality, Precommitment, and Constraints*. Cambridge: Cambridge University Press.

Evans, K. 2002. Taking control of their lives? Agency in young adult transitions in England and the New Germany. *Journal of Youth Studies* 5(3): 245–69.

Fennell, S. 2020. Youth employment, informality, and precarity in the Global South. In S. Swartz, A. Cooper, C. M. Batan and L. K. Causa (eds), *The Oxford Handbook of Global South Youth Studies*. Oxford: Oxford University Press, 286–303.

Furlong, A. 2015. Transitions, cultures, and identities: What is youth studies? In D. Woodman and A. Bennett (eds), *Youth Cultures, Transitions, and Generations: Bridging the Gap in Youth Research*. London: Palgrave Macmillan, 16–27.

Galtung, J. 1969. Violence, peace, and peace research. *Journal of Peace Research* 6(3): 167–91.

Giddens, A. 1986. *The Constitution of Society: Outline of the Theory of Structuration*. Berkeley, CA: University of California Press.

Hoffman, D. 2011. Violence, just in time: War and work in contemporary West Africa. *Cultural Anthropology* 26(1): 34–57.

Hutchinson, F. P. and Inayatullah, S. 2010. Futures studies and peace studies. *The Oxford International Encyclopedia of Peace*, vol. 2. Oxford: Oxford University Press, 174–80.

ILO (International Labour Organization). 2014. *Liberia Youth Employment Skills Project and Youth Employment Program: Country Pilot Report by the Interagency Social Protection Assessment Initiative*. Available at https://www.ilo.org/wcmsp5/groups/public/---ed_emp/documents/projectdocumentation/wcms_504717.pdf (accessed 27 December 2023).

Inayatullah, S. 2002. Youth dissent: Multiple perspectives on youth futures. In In J. Gidley and S. Inayatullah (eds), *Youth Futures: Comparative Research and Transformative Visions*. Santa Barbara, CA: Greenwood Publishing Group, 19–30.

Jenkins, J., Mokuwa, E. Y., Peters, K. and Richards, P. 2020. Changing women's lives and livelihoods: Motorcycle taxis in rural Liberia and Sierra Leone. *Proceedings of the Institution of Civil Engineers-Transport* 173(2): 132–43.

Johnston-Goodstar, K. 2020. Decolonizing youth development: Re-imagining youthwork for Indigenous youth futures. *AlterNative: An International Journal of Indigenous Peoples* 16(4): 378–86.

Koskimaki, L., 2017. Youth futures and a masculine development ethos in the regional story of Uttarakhand. *Journal of South Asian Development* 12(2): 136–54.

Leonardsson, H. and Rudd, G. 2018. The 'local turn' in peacebuilding: A literature review of effective and emancipatory local peacebuilding. In J. Ojendal, I. Schierenbeck and C. Hughes (eds), *The 'Local Turn' in Peacebuilding: The Liberal Peace Challenged*. London: Routledge, 9–23.

MacDonald, R. 2009. Precarious work: Risk, choice and poverty traps. In A. Furlong (ed.), *Handbook of Youth and Young Adulthood*. London: Routledge, 183–91.

MacDonald, R. and King, H. 2021. Looking south: What can youth studies in the Global North learn from research on youth and policy in the Middle East and North African countries? *Mediterranean Politics* 26(3): 285–307.

Mac Ginty, R. 2021. *Everyday Peace: How So-called Ordinary People Can Disrupt Violent Conflict*. Oxford: Oxford University Press.

Mac Ginty, R. and Richmond, O. P. 2013. The local turn in peace building: A critical agenda for peace. *Third World Quarterly* 34(5): 763–83.

McMullin, J. R. 2022. Hustling, cycling, peacebuilding: Narrating postwar reintegration through livelihood in Liberia. *Review of International Studies* 48(1): 67–90.

Munive, J. 2010. The army of 'unemployed' young people. *Young* 18(3): 321–38.

Page, J. 2013. For Africa's youth, jobs are job one. In *Foresight Africa: Top Priorities for the Continent in 2013*. Washington, DC: Brookings Institution. Available at https://www.brookings.edu/articles/foresight-africa-top-priorities-for-the-continent-in-2013/ (accessed 27 December 2023).

Peeters, P. 2009. *Youth Employment in Sierra Leone: Sustainable Livelihood Opportunities in a Post-Conflict Setting*. Washington, DC: World Bank Publications.

Podder, S. 2015. The power in-between: Youth's subaltern agency and the post-conflict everyday. *Peacebuilding* 3(1): 36–57.

Podder, S. 2022. *Peacebuilding Legacy: Programming for Change and Young People's Attitudes to Peace*. Oxford: Oxford University Press.

Podder, S. 2023. Land back to the people or not?: The variable pathways of civic mobilisation against land grabs in rural Sierra Leone. *Journal of Modern African Studies*.

Podder, S., Prelis, S. and Sankaituah, J. 2021. Youth, peace and programming for change: Critical reflection between the university, the headquarters and the field. *Global Policy*. Available at https://www.globalpolicyjournal.com/sites/default/files/pdf/Podder%2C%20Prelis%20and%20Sankaituah%20-%20Youth%2C%20Peace%20and%20Programming%20for%20Change.pdf (accessed 27 December 2023).

Pospisil, J. 2019. *Peace in Political Unsettlement: Beyond Solving Conflict*. London: Palgrave Macmillan.

Roemer, J. E. 1998. *Equality of Opportunity*. Cambridge, MA: Harvard University Press.

Sen, A. 1999. *Development as Freedom*. New York: Oxford University Press.

Simon, H. A. 1990. Bounded rationality. In J. Eatwell, M. Milgate and P. Newman (eds), *Utility and Probability*. London: Palgrave Macmillan, 15–18.

Sommers, M. 2011. Governance, security and culture: Assessing Africa's youth bulge. *International Journal of Conflict and Violence* 5(2): 292–303.

Stewart, F. 2009. *Policies Towards Horizontal Inequalities in Post-Conflict Reconstruction*. London: Palgrave Macmillan, 136–74.

Sumberg, J., Flynn, J., Mader, P., Mwaura, G., Oosterom, M., Sam-Kpakra, R. and Shittu, A. I. 2019. Formal-sector employment and Africa's youth employment crisis: Irrelevance or policy priority? *Development Policy Review* 38(4): 428–40.

UNDP Liberia. 2023. *Livelihood and Employment Creation Project*. Available at https://www.undp.org/liberia/projects/livelihood-and-employment-creation-project (accessed 27 December 2023).

UNDP Sierra Leone. 2011. *Youth Employment and Empowerment Programme* [Brochure]. Freetown: UNDP.

UNDP Sierra Leone. 2016. *Youth Entrepreneurship and Employment Project 3.0: 2016–2018 Programme Document*. Freetown: UNDP.

Winter, J. M. 2006. *Dreams of Peace and Freedom: Utopian Moments in the Twentieth Century*. New Haven, CT: Yale University Press.

Wood, E. J. 2008. The social processes of civil war: The wartime transformation of social networks. *Annual Review of Political Science* 11: 539–61.

World Bank. 2017. *Africa's Pulse*. 16. Washington, DC: World Bank Group.

# 4

# Visualising Hope in the Radical Data Work of W. E. B. Du Bois

*Kiran K. Phull*

## Introduction

At the 1900 Exposition Universelle in Paris, France, where an audience of close to fifty million spectators came to revel in the pageantry of an emergent modern world on display, the American theorist and activist W. E. B. Du Bois was invited to contribute an exhibit on historical, sociological and intellectual aspects of Black life in the United States. This occurred at a critical moment in global race relations, coinciding with the First Pan-African Conference in London where Du Bois expounded on the problem of the global colour line. The 'radicalisation of black politics' gained expression through the reflexive sociological and statistical work that he later displayed in Paris (Bini 2014, 40). In the setting of a colonial metropolis, this subversive data-driven exhibit sought to empower the racialised subject and to educate broader Western audiences, opening a social space for African Americans to engage in transnational dialogues of social reform (Bini 2014). A unique feature of the exhibit was a mounted catalogue of over sixty large-format charts, maps and data visualisations hand-crafted by Du Bois with the assistance of students from Atlanta University. An exemplary case of *data as craft*, these 'data portraits' traced movements and improvements in occupation, education, literacy and other social factors in the decades following the abolition of slavery (Battle-Baptiste and Rusert 2018). Donated to the Library of Congress soon after the Paris Exposition, the visualisations remained largely hidden from view for over a century until they were digitised in 2014. Since then, Du Bois's data portraits have captivated the social imaginary and awakened a more inclusive dialogue around the possibilities for data visualisation as a means towards liberation and justice (Fusco and Olman 2021).

This chapter revisits Du Bois's visual statistical work in order to understand how hope operates through his data practice, and what this might mean for the possibility of hope in a data-driven world. In an effort to confront the violence of statistical abstraction that threatens to reduce the expression of human life to classifications, calculations, regularities and evidence, the chapter opens with a reflection on the possibilities for statistical hope through a consideration of Du Bois's own understanding of hope and the lessons of his generative visual praxis. Taking these visualisations as a site of critical exploration, the chapter argues that hope is materially and figuratively encoded as an essential, tacit variable. Drawing on three examples that foreground conditions of the local, the state and the international, I find that hope in the Du Boisian statistical imagination, as 'a faith in the ultimate justice of things' (Du Bois 2009, 175), is tinged with complexity. In portraying a hope that diagrammatically confronts the burden of history, the present struggle and the expectant future, Du Bois invites critical reflection on the necessarily conflicted nature of hope and possibility in a world afflicted by inequality and racial violence. Within these visualisations, an active dialectic of hope and suffering emerges, where the possibilities for self-realisation and self-representation collide with the material conditions of the global colour line in vivid graphs, charts, maps and other diagrams. Here, I consider Du Bois's method of translating statistical thinking into more profound narrative form, which remains an essential contribution to visual politics and critical international thought today. The chapter concludes with a reflection on the possibilities for hopeful data futures, arguing that the ethics and aesthetics of Du Bois's encoded hope can help us to think about how data in a difficult world is not incompatible with crafting a better one.

## Locating a Du Boisian hope

A tempered sense of hope appears to different degrees in readings of Du Bois. Joseph R. Winters explores a melancholic expression of hope through the trope of sorrow in *The Souls of Black Folk*, where experiences from within the Veil, as an organising principle of racialised existence, inhibit progress and the flourishing of Black life (Winters 2016). Hope here is shaded by the isolating conditions of the colour line, which render the living world 'strange' and 'crooked' (Winters 2016, 69). Reflecting in *Souls* on the death of his infant son from diphtheria, Du Bois wrote of 'a hope not hopeless but unhopeful' (Du Bois 2009, 141) where life's fragility extends to a future that is contingent upon sorrow and suffering. But as Winters shows, there remain 'flickering moments of hope, possibility, and transcendence' situated between death and liberation, between present and future (2016, 74). As Nahum Dimitri Chandler suggests, this 'unhopeful hope' is distinctly radical for its 'acceptance without resignation' of the ontological constraints of the Veil (2022, 129). Similarly, the introduction to this volume describes a nihilist, deconstructive hope that enables other modes of existence beyond this world. The 'scalar shifts in vantage point' that position Du Bois 'outside modernity' are precisely what enables his theorising from within the Veil (Fusco and Olman 2021, 174).

Hope in the Du Boisian sense also finds itself intrinsically linked to progress and ideas about social advancement. Joe P. L. Davidson argues that Du Bois's narrative writing serves to critically challenge the dominant notions of progress that saturated nineteenth- and early twentieth-century social theory; in particular, the belief in the linear trajectory of human development as a defining feature of modernity (2021). By offering a more nuanced understanding of temporality, 'Du Bois rejects the idea that history moves in a necessary and ameliorative fashion towards an ever better world' (Davidson 2021, 382). In Davidson's reading, Du Bois's idea of progress is expressed through a latent power that seeks to overcome 'the catastrophic processes of American history', legacies of slavery and racial violence that enact barriers to the self-realisation of Black folk (2021, 388). Progress, refracted through a clouded view towards the future, is necessarily ugly and ruinous because hope for a better world is always mediated by the disappointment that attaches itself to the struggle of the Black experience.

In a similar way, Mark S. Cladis contends with questions of resiliency and catastrophe in examining Du Bois's notion of a 'dark, wild hope' (2020). At the nexus of political and environmental racism, Cladis locates a sense of hope that is tied to the wild beauty and resistance of nature. The aesthetics of Du Bois's written narratives relate 'the profound, intertwined racist and environmental oppression of people, place, and land', where resistance and hope inscribe themselves into the struggle against exploitative racist and capitalist practices (Cladis 2020, 218). Holding hope for a more promising future is still necessary if we are to overcome the looming crises of climate change, capitalist exploitation and white supremacism, but this hope must lean into the grief and loss that has been sustained in order to become something transformative. This expression of hope is decidedly unoptimistic (see Waldow, Bargués and Chandler in the introduction to this volume). It is also more than an affect, emotion or desire. For Cladis, hope must be understood as practice – as a vocation that demands our intention and propels our resiliency against struggle.

This chapter extends the perspectives above in observing how the dialectic of hope and suffering within Du Bois's intellectual work finds its way into his statistical creations. It takes seriously Cladis's notion of *hope as practice* in reflecting upon the ways by which Du Bois's generative visual praxis produces an expansive material-aesthetic view of hope. Here, I argue that Du Bois's material-aesthetic vision of the world is situated at the critical demarcation of the global colour line. This form of hope is agentic but stops short of idealising a world of perfect equality. Conceived at an earlier stage in his academic career before becoming disillusioned with 'the ivory tower of statistics and investigation' (Du Bois 2007, 111), this hope is perhaps more expansive than what we find in Du Bois's later narrative works. In his statistical imaginary, hope moves along a continuum from the bleak to the expectant, where the future – still mediated by suffering – holds active potential for radical progress.

In sourcing different configurations of hope within these visualisations, the chapter asks what it might mean more broadly to craft data as hope. Exploring this question, the chapter brings together critical readings on the symbolic and emancipatory potential of Du Bois's visual statistical work (Bering-Porter 2022; Murphy 2021; Weheliye, 2015). It also unites emerging interdisciplinary perspectives that analyse and contextualise the visualisations and their import for contemporary social theorising (Battle-Baptiste and Rusert 2018; Bini 2014; Fusco and Olman 2021; Olman 2022), bringing these into conversation with debates within international studies exploring the now inextricable and constitutive relationship between data and power (Aradau and Huysmans 2014) and the 'craft' of worldmaking (Austin and Leander 2021; Getachew 2019). The following section explores Du Boisian statistical hope at the levels of the local, the state and the international, which helps us to think about the ways by which Du Bois's method invites possibilities for theorising the world differently.

## Spatial Configurations of Hope in Du Bois's Data Portraits

In 1900 Du Bois travelled to Paris to take part in the cosmopolitan spectacle of the World Fair and present a scientific catalogue of 'subversive visual innovations' that portrayed the conditions of the global colour line in vivid graphical form (Olman 2022, 155). As a show of *fin de siècle* imperial mastery, modernity and progress, the World Fair was staged to usher in a new century where ideas of racial determinism remained firmly entrenched. It was in this setting that Du Bois's sociological work sought to position Black subjecthood as 'integral to the formation of modernity' (Wilson 2012, 107). Du Bois, an Atlanta University scholar at the time, was asked to contribute to a generative exposition of Black life and progress on the world stage, envisioned as a composite of sociological analysis, photographs, documentation and more. Du Bois's key contributions included a recently published ethnographic study (*The Philadelphia Negro*) in addition to an empirical study of Georgia's Black communities, for which he was awarded a stipend to hire a group of students to assist in its production. Using census data and inductive survey methods, Du Bois and his team produced over sixty hand-drafted charts and graphs for two series, the first being 'The Georgia Negro: A Social Study' (comprised of thirty-six plates) and the second titled 'A Series of Statistical Charts Illustrating the Condition of the Descendants of Former African Slaves Now Resident in the United States of America' (comprised of twenty-six plates) (Figure 4.1). Debuted to a mainly European audience, the exhibit won a Grand Prix and Du Bois was awarded a gold medal for the graphical work.

Each of the hand-crafted charts, graphs and statistical plates contained within Du Bois's catalogue is characterised by saturated colours, bold typeface and a modernist aesthetic quality. Their uniform starkness – at once beautifully abstracted and visually restrained – is particularly striking, while the interplay of form, plane and geometric configuration animates the otherwise negative spatial field. The visualisations were displayed in a corner section in the Palace of Social Economy and Congress – a visionary and unprecedented portrayal of subversive data through the art of statistical portraiture. Aesthetically, the visualisations were both informed by and departed from other forms of data visualisation seen at the time. While methods for visualising data were growing in close collaboration with the statistical imaginary of modernity, the visual medium allowed for a strategic encoding of resistance, taking seriously 'the matter of *colour* and *line* as bearing not just metaphorical but textual, material, and representational import for Du Bois' (Fusco and Olman 2021, 161). Crucially, each of Du Bois's visualisations tells us something of the paradox of hope, where the possibilities for a more just and equitable world remain tied to the onto-epistemic regime of racial difference that sustains this one.

In this section, I consider the encoding of hope through the material and transcendent design of data, focusing on three visualisations that embed radical possibility at the level of the local, the state and the international. Each invokes an agentic hope that stops short of idealism but opens pathways to futurity. Each embodies historic, present and future conditions of modern Black subjecthood, enabling 'the proximate visualisation of abstract forces of power rather than the representation of individuals or groups' (Weheliye 2015, 26). Contrary to conventional statistical objectives like determining reductive patterns, describing data distributions or architecting classificatory schemata, Du Bois's generative approach to statistical visualisation achieves two aims. The first is the representation of the subject of knowledge *as itself* and not conscripted into a broader schema of racial and social ordering. The second is the distinct ability of the visualisations to generate meaning beyond the canvas; they can be read both as narratives of provocation and as counter-historical

**Figure 4.1** Photograph of the African American exhibit at the Paris Exposition (November 1900). Library of Congress, LC-DIG-ppmsc-04826.

archives of the past, present and future. In this way, Du Bois's methodological approach stands as an explicit statement of agentic possibility through visual data-crafting, counteracting the persistent violence of statistical abstraction.

A first example below takes the 'Du Boisian spiral' as emblematic of the progressive yet non-linear trajectory of Black life at the turn of the twentieth century. Foregrounding the demographics of the local, Figure 4.2, from Du Bois's 1900 Georgia study, depicts the total numbers of large and small city- and rural-dwelling Black Georgians in 1890 in one continuous yet visually interrupted line. As a temporal snapshot, the figure tells us that *fin de siècle* Black Georgia was predominately rural, represented by the captivating red spiral that that draws the viewer's gaze and travels upward towards the broken geometric route of city-dwellers.

Here, the rigid linearity of a standard bar graph is translated into a mesmeric helix, challenging the reductive uniformity of quantitative methods that remains so firmly entrenched in the empirical social sciences. Indeed, the recurring motif of the spiral in Du Bois's diagrams works both with and against linearity, inviting playful circular and spatial momentum to the fixed nature of data. If we were to imagine changes to the demographic fabric of Georgia that ensued in the decades after 1890, we would witness a gradual unravelling of the central spiral, as urban hubs like Atlanta and Savannah drew increasingly from rural population reserves. As if the spiral holds kinetic potential, we imagine revolutions in motion through time. Replete with the possibility of change and transformation, the diagram draws the viewer's awareness towards a future not pictured on the canvas – that of the helical uplift of Black life post-Emanciaption.

Aesthetically, Du Bois's spirals have been described as beautiful, disorienting, absurdist, glyphic, and as something that 'defies categorisation' (Battle-Baptiste and Rusert 2018). Cyclical vortices and straight lines that turn back on themselves appear throughout his visual statistical work. Employing a method that transforms data into narrative form, this approach challenges teleological visions of civilisational history and rejects a unidirectional course for racial and social progress. Progress, in the Du Boisian statistical sense, is not linear but 'conflicted, contingent, and reversible' as it is inseparable from the broader structures of inequality that permeate history (Ray et al. 2017, 148). Tethered to the global colour line, this conflicted progress enables change but remains mediated by discontent – it is not hopeless. For Du Bois, the forward motion of progress 'with all its retrogression is a spiral not a circle, and as long as there is motion there is hope' (in Chandler 2022, 187).

Rising above the spiral in Figure 4.2 is the 'brute might' of modern urban life (Du Bois 1920, 237). Here, a horizontal line forms a 'critical graphical

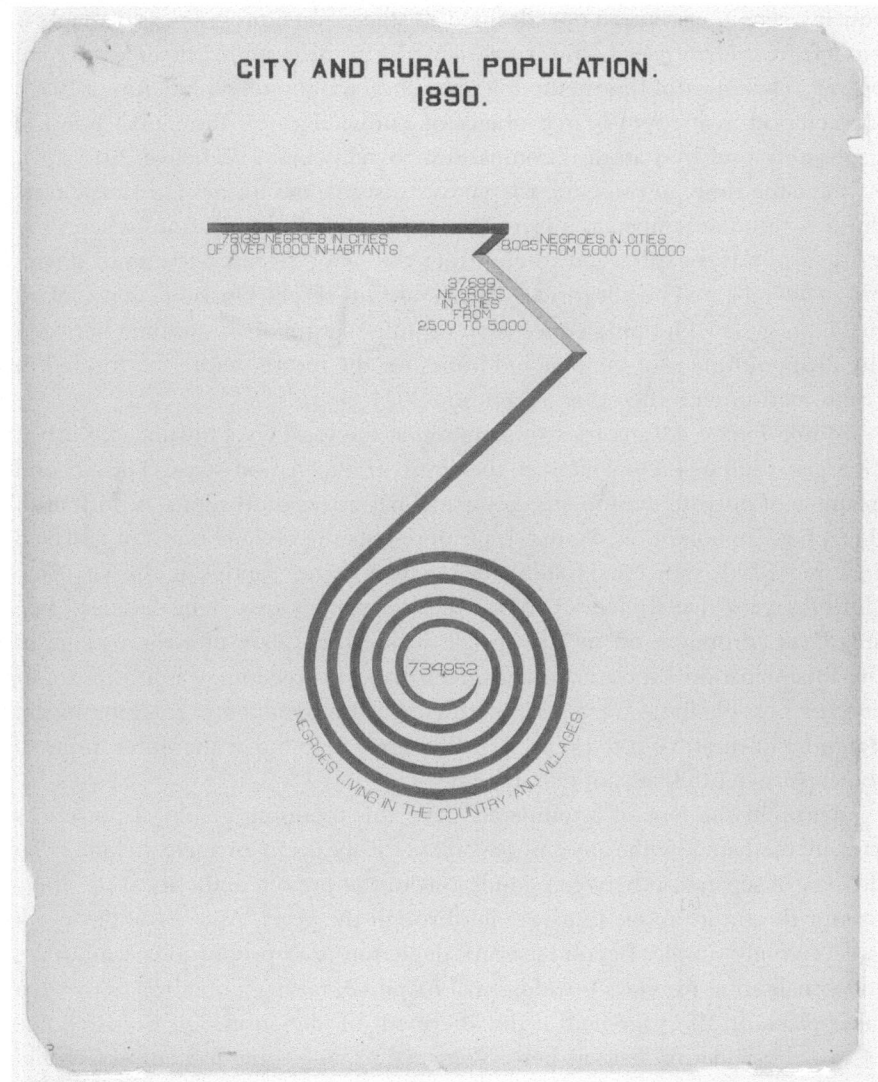

**Figure 4.2** 'City and Rural Population, 1890', W. E. B. Du Bois (1900). Library of Congress, LC-DIG-ppmsca-33873 [Plate 11].

horizon' of possibility for another world (Weheliye 2015, 40), which distinctly orients the viewer's gaze towards the future. The graphical techniques by which Du Bois maps temporal horizons lead us to ask what kinds of futures are possible within his statistical imaginary. The actualisation of racial equality in Du Bois's data work remains somewhat ambivalent. The question of racial equality is neither centred nor idealised. Rather, Black progress as a variable is often treated *independently* and *in spite of* white positioning, status and structural power. The effect of this methodological choice is the representation of Black subjecthood as its own 'stated object of knowledge' on the global political stage, unbound by statistical comparison to whiteness (Weheliye 2015, 24). At the same time, in orienting the viewer towards the future, Du Bois's anti-linear technique tempers the prospects of life after Emancipation, where 'the vestiges of slavery still acted to construct the scope of black freedom' (Hartman 1997, 172). The allegorical data spiral thus recalls Du Bois's sense of an 'ugly' progress: 'a looping conception of time that involves shuffling between the disappointment of the past and hopes for the future, with each formed in confrontation with the other' (Davidson 2021, 383).

While Figure 4.2 invites consideration at the local level in rural and urban Georgia, Figure 4.3 operates at the level of the nation-state. This second example of hopeful tension engages spatial narratives of freedom. A portrait of the American condition, Figure 4.3 (re)presents the divided status of enslaved and free Black subjects. Holding the viewer's gaze captive is the jet-black cliff of slavery that stretches towards the right and across eight decades of an area chart, dropping off in a steep fall during the 1860s after the passage of the Emancipation Proclamation. A green band of freedom – as a measure of another possible life – wavers across the upper horizon before reaching its full potential in the final data point. Visually, hope is found at the outer limits of the expansive black area of entrapment.

Through the lens of a tempered hope this totalising graphical portrait is visually mediated by the abyss of slavery, yet it speaks to emancipated life. The degrees of separation between conditions of the present and critical possibilities for the future move transversally through the years. As a visual device, it is deceivingly simple. Beyond its stark depiction of disproportionate injustice, the visualisation provides no additional narrative, raising a question as to what takes place off the page and in the aftermath of liberation. Du Bois's choice of spatial orientation matters here. Were we to invert the diagram, we would encounter a different, more optimistic narrative of emancipation – one that depicts progress as a sharp upward trajectory of the freedom line, ascending towards a higher state of being-in-the-world. But in this rendering, progress is a falling line of flight that seeks to escape the present condition. With this, Du Bois leaves the future both undescribed and unimaginable.

This negative condition of possibility remains deeply agentic. The visual metaphor allows for the possibility of theorising from within the black abyss. Murphy argues that the looming black edifice of slavery is a product of chance, where structural conditions of oppression were sustained by 'innumerable indeterminate human acts' of control and 'of human decisions that have no

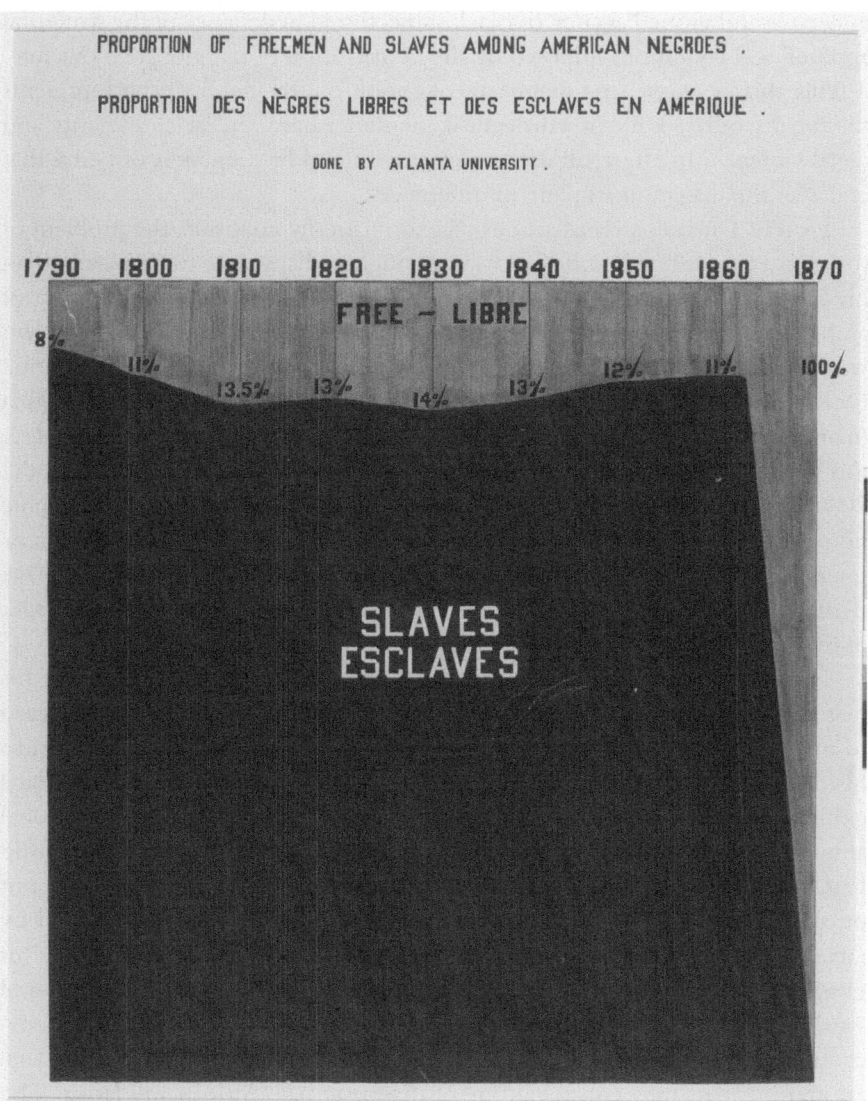

**Figure 4.3** 'Proportion of Freemen and Slaves Among American Negroes', W. E. B. Du Bois (1900). Library of Congress, LC-DIG-ppmsca-33913 [Plate 51].

natural or inevitable foundation' (2021, 214). At the same time, freedom 'alters the landscape' of the diagram and makes visible the incalculable role of agency as a crucial mechanism behind the spectacular fall of the system of slavery (Murphy 2021, 214). Competing agencies thus operate on either side of the colour line. In place of this dichotomy, we might also read the diagram as the collective and unified agency of Black subjecthood in defiance of the structural and political conditions imposed by the demarcation of the global colour line. Taking this view, we find that each data point on the graph always represents the totality of Black life in a dialectical encounter between racial solidarity and racial violence. In effect, all data points are unified by a oneness of being that collapses the diagrammatic line of difference.

Each of Du Bois's visualisations diagrammatically confronts the problem of the global colour line in different ways, though perhaps none more plainly than his cartographic reproductions. In Figure 4.4, a hand-rendered illustration of global hierarchies of difference is transposed upon a Mercator projection of two worlds 'split in half' and conjoined by power lines that dash across the Atlantic Ocean, tethering the African continent to the Americas and the state of Georgia (marked with a star) (Olman 2022; Wilson 2018). Etched prophetically below this stark and synoptic visual is the central proclamation: 'The problem of the 20$^{th}$ century is the problem of the colour-line'. The image traces a cartographic history of the Black Atlantic slave trade, which 'allows Du Bois to visually represent hundreds of years and thousands of miles of oppression' in a single image (Battle-Baptiste and Rusert 2018). The map thus becomes immeasurably larger in meaning than what it can contain.

In this and other maps produced for the exhibition, 'Du Bois and his team redeployed the western methods of cartography that had been used to marginalise and exploit black life by inscribing the black world back into history and geography' (Wilson 2018, 42). As the opening plate to the Paris exhibit, the Black Atlantic world map draws strong links between global anti-colonial and local anti-racist struggles. In the setting of the 1900 World Fair where the dawning of a new century of progress was theatrically put on show, Du Bois's map presages a century of global conflict governed by racial, ethnic and hierarchical logics. At the level of the international, his visual praxis resists white supremacist ideals of colonial order and imperial worldmaking (Anievas et al. 2014). In doing so, it foreshadows twentieth-century decolonial struggles and successes, while the map's lines of bondage are transformed into constellations of diasporic pan-African solidarity. The subversive remapping of history and future makes visible the political lines that stratify the international. This atlas of inequity demarcates 'the abstract forces of the colour line, while simultaneously putting under erasure its primary function, which is to create and maintain hierarchical caesuras between different groups of humans, especially between black and white subjects' (Weheliye 2015, 39). As a dissenting narrative of the world order, Du Bois's cartographic praxis allows for the theorising of global power relations from a different vantage point.

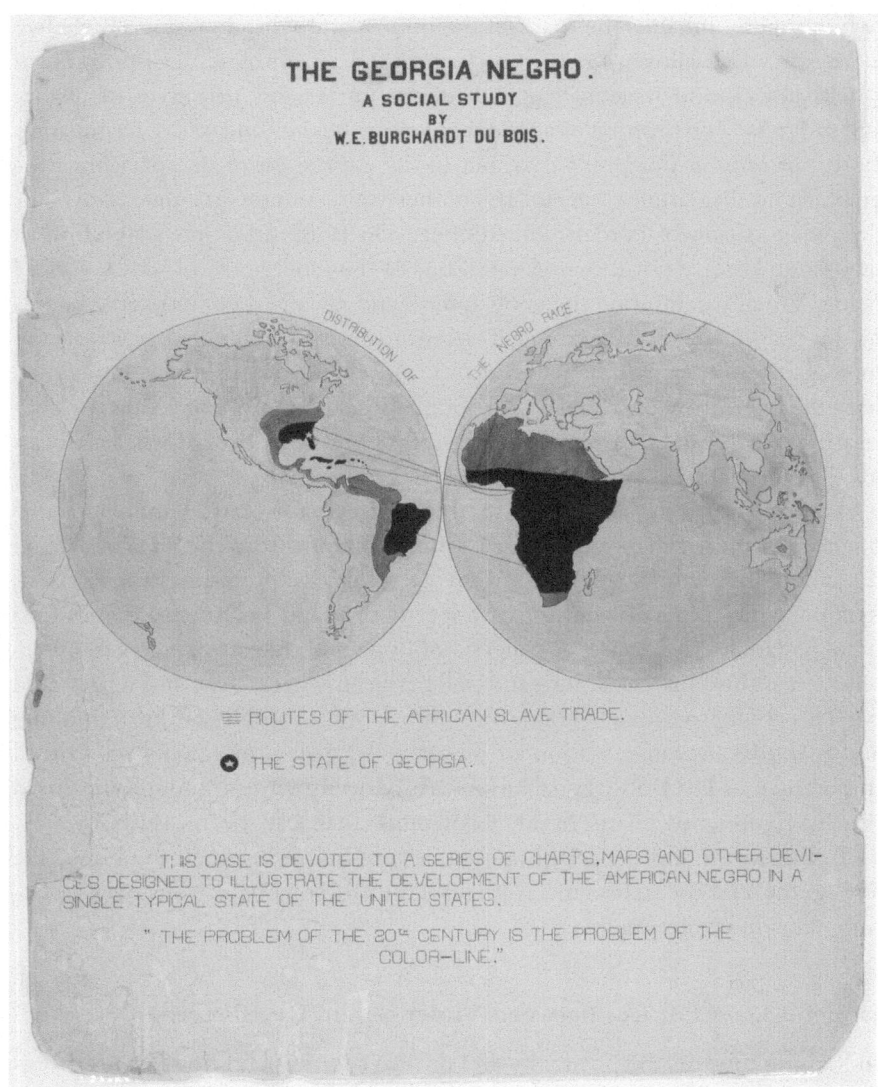

Figure 4.4 *The Georgia Negro: A Social Study*, W. E. B. Du Bois. Diagram shows routes of the African slave trade with the state of Georgia starred (1900). Library of Congress, LC-DIG-ppmsca-33863 [Plate 1].

This form of counter-mapping that experiments 'with method in ways that enact worlds differently' (Aradau and Huysmans 2014, 608) makes space for the potential to encode hope. In one way, this counter-mapping renders the political visible, exposing the hidden modalities of power that seek to manage 'who properly inhabits the space of the human and who does not' (Weheliye 2015, 27). This allows Du Bois to foreground an ontological disposition of Black subjecthood that challenges Western universalist narratives of history and progress. The promise of encoding identity, agency and self-determination exists not only in Du Bois's data, but in the expression of that data in cartographic and diagrammatic form. In another way, counter-mapping allows for the renegotiation of borders and frontiers. Du Bois's atlas draws attention to new boundaries – not those of states, but of the emergence of states of non/being. Visually delimiting the geographical and racialised conditions of being-in-the-world, the map centres self-determination at a moment between the inescapable history of colonial violence and an unwritten future concerned with the overthrow of white-world domination. This moment recalls the necessarily 'ugly' character of hope: a sense of progress that evokes what has been lost (Davidson 2021).

Returning to the metaphor of the horizon, this cartographic practice operates in the service of hope for a world different from the present. While the colour line works to erase and divide, it also delimits an event horizon, beyond which 'the orientation of hope, desire, and becoming' comes into view (Chandler 2022, 44). A dialectic of hope and despair operates transversally across this horizon; a push and pull between what is now and what could be. This ability to move beyond the confines of descriptive statistical thinking and towards the renegotiation of global racial orders highlights the critical importance of Du Bois's visualisations. As Murphy writes, 'though measured out by figures, his charts figure forth more than can be measured' (2021, 215). Figurative and material hope, on a continuum between the bleak, the ambiguous and the expectant, is preserved in these visualisations over a century after they were made.

## Hopeful Data Futures between Violence and Resistance

In looking towards the possibility and prospect for hopeful data futures, I suggest that Du Bois's method of encoding hope in his visual statistical work – a hope 'not hopeless but unhopeful' (Du Bois 2009, 141) – might help us to think about how data in a difficult world is not incompatible with crafting a better one. In this section, I consider three ways in which Du Bois's method works to confront violent data practices of exploitation and expropriation, offering a productive challenge to the datafication of contemporary life (Bering-Porter 2022). These include overcoming the violence of statistical abstraction through humanistic quantification, enabling subjecthood through visual narratives, and imagining the world through aesthetic practice.

## Countering the Violence of Abstraction

Counting, classifying and categorising human life has long been used to stratify people on the basis of essentialised differences, which in turn sustain global racial hierarchies. Why, then, does Du Bois resort to the methods of generalisation and visual abstraction to subvert this same violence? Contending with this paradox of employing quantification to overcome its very constraints, Murphy (2021) argues that Du Bois's statistical method is in actuality a metaphorical practice, which substitutes the violence of abstraction for generative meaning. Countering dehumanising statistics need not dispose of quantification altogether. Rather, Du Bois relies on a form of visual metonymy that moves data beyond numbers and towards concepts, theories and affective states of being. The ability to craft narrative and enact worlds through numbers emerges from a statistical episteme that reaches beyond methods of calculation and quantification and can instead be located in 'the veiled world beyond' (Du Bois 2009, 200).

Thinking another way, Weheliye highlights the intentionality behind Du Bois's use of descriptive statistics, which 'hesitate before the rules of generality by refusing to infer immutable laws', rejecting direct comparison, and thus giving space to critically reflect upon and describe the world (2015, 27). Quantification thus becomes a tool of worldmaking, and it is here that hope is manifested. While the visualisations form part of the early corpus of Du Bois's sociological work, created before he grew critical of the social scientific ethos of detachment and objectivity and abandoned the ivory tower of statistical investigation, they nevertheless retain a disruptive potential. Statistics, in the Du Boisian sense, is therefore more than a mathematical tool with analytical utility – it is a visual metonymic form of expression charged with meaning.

## Visualising Subjecthood

Can subjecthood be retrieved from quantified methods when the unsympathetic calculations of statistical treatment risk dehumanising subjects? The image of a cold, detached statistician casts a shadow over more than a century of empirical social scientific endeavouring. In Saidiya Hartman's words, statistics are not bloodless but are antagonists in the tragedy of racial violence (2019). Hartman narrates how the transformation of women, mothers and husbands into tables and ratios rendered them small and insignificant, as cold statisticians 'transposed lives into bars and curves, densities of ink and colour' (2019, 105). Similary, Tukufu Zuberi recounts the deleterious effects of the statistical analysis of race, where observational records of racialised lives reduced people to objects of statistical, biological and demographic control (2001). Acknowledging the violence of racial classification in his 1900 address 'To the Nations of the World', Du Bois extended the problem of the colour line to the question of how far differences of race would form 'the basis of denying to over half the world the right of sharing to their utmost ability the opportunities and privileges of modern civilisation' (Du Bois 2022).

In employing methods for quantifying racialised bodies, Du Bois returns subjecthood to his data in critical ways. Weheliye argues that his use of statistics and diagrams works to 'desediment' the Black subject from biological and phenotypical representations, instead historicising and repositioning his subjects as a category that comes about through the violent imposition of the colour line (2015, 27). In visualising the historically contingent reality of racialised life, Du Bois enables his subjects to be understood in a new light while also encouraging his audience to see beyond the limits of their knowledge. Visualisation, for Du Bois, is thus a form of noticing and a method of self-representation. Importantly, the representation of his data subjects is never static but always in motion – bound up in spatial and temporal relations with historical and future time. Setting his subjects in motion, the visualisations propel the forward momentum of Black life 'resolutely toward the future' (Du Bois 2009, 55). In this way, Du Bois offers the possibility for embedding agentic subjecthood in statistical work.

## *The Aesthetics of Imagining*

How does Du Bois's aesthetic practice enact alternate visions of the world? The aesthetic hallmarks of his visualisations have been lauded for their stylistic and rhetorical prescience (Bering-Porter 2022). They predate the European modernist movements of twentieth-century fine art and yet evoke the universal abstractions of Mondrian, Malevich and Kandinsky. They illustrate racialised conditions through quantification yet anticipate the contemporary 'cultural imaginary that dreams of a world of data' (Bering-Porter 2022, 269). The sensory cues of vibrant colour, structural rigidity, counter-linearity and spatial experimentation stand in for belaboured enumeration, description or textual explanation. They express a stark and visceral power that, in the words of Battle-Baptiste and Rusert, defines the aesthetics of the colour line (2018).

The effect is a visual subversion of the racial violence that finds itself embedded in white supremacist notions of modernity, science and civilisation. Coloured lines become conceptual strands of reasoning that allow the social and political world to be theorised from a different vantage point. Thinking of data as a kind of imaginary medium, Bering-Porter suggests that Du Bois's visual praxis 'makes it possible to see the intricate structures and unseen connections that tie the world together through causal relations that are otherwise invisible' (2022, 278). The colour line is more than a site of theoretical exploration; in visual form it becomes visceral, experiential and totalising. Visualisation thus has a speculative function that is externalised through the material-aesthetic practice of crafting data (Murphy 2021). With this method, Du Bois's visual statistical imaginary is productive of alternative visions of the world – as a generative and expectantly hopeful act of anti-colonial worldmaking (Getachew 2019).

## Conclusion

While the global colour line as a research agenda has emerged as a productive area of critical and postcolonial international theorising, Du Bois's visual statistical work has so far garnered less attention within international studies. This chapter examines his data portraiture through the lens of a tentative hope, arguing that an expansive dialectic of hope and suffering is encoded as an essential and tacit feature of his method. Hope, in this light, is agentic. More than an affect, it underpins the material-aesthetic practice of crafting data as a means towards liberation and justice. As a practice, it seeks to make visible the ways by which the present struggle is always contingent upon past suffering. At the same time, it holds space for the possibility of a future world that is neither idealistic nor optimistic but seeks to disrupt and dismantle the confines of the present one. Connecting local anti-racist with global anti-colonial struggles, Du Bois's data portraits transcend the basic utility of data visualisation. In this moment where the inextricable and constitutive relations between data and power stand to dehumanise and disempower, Du Bois's method offers a possibility for emancipatory and hopeful data futures.

## References

Anievas, A., Manchanda, N. and Shilliam, R. 2014. Confronting the global colour line: An introduction. In A. Anievas, N., Manchanda and R. Shilliam (eds), *Race and Racism in International Relations: Confronting the Global Colour Line*. Abingdon: Routledge, 1–15.

Aradau, C. and Huysmans, J. 2014. Critical methods in international relations: The politics of techniques, devices and acts. *European Journal of International Relations* 20(3): 596–619.

Austin, J. L. and Leander, A. 2021. Designing-with/in world politics: Manifestos for an international political design. *Political Anthropological Research on International Social Sciences (PARISS)* 2: 83–154.

Battle-Baptiste, W. and Rusert, B. (eds). 2018. *W. E. B. Du Bois's Data Portraits: Visualizing Black America*. New York: Princeton Architectural Press.

Bering-Porter, D. 2022. Data as symbolic form: Datafication and the imaginary media of W. E. B. Du Bois. *Critical Inquiry* 48(2): 262–85.

Bini, E. 2014. Drawing a global color line: 'The American Negro Exhibit' at the 1900 Paris Exposition. In G. Abbattista (ed.), *Moving Bodies, Displaying Nations. National Cultures, Race and Gender in World Expositions 19th to 21st Century*. Trieste: EUT Edizioni Università di Trieste, 39–65.

Chandler, N. D. 2022. *'Beyond This Narrow Now': Or, Delimitations, of W. E. B. Du Bois*. Durham, NC: Duke University Press.

Cladis, M. S. 2020. Du Bois and dark, wild hope in an age of environmental and political catastrophe. *Ecozon@: European Journal of Literature, Culture and Environment* 11(2): 216–23.

Davidson, J. P. L. 2021. Ugly progress: W. E. B. Du Bois's sociology of the future. *The Sociological Review* 69(2): 382–95.

Du Bois, W. E. B. 1920. *Darkwater: Voices from within the Veil*. New York: Harcourt, Brace and Howe.

Du Bois, W. E. B. 2007. *Dusk of Dawn: An Essay Toward an Autobiography of a Race Concept*. H. L. Gates Jr (ed.). Oxford: Oxford University Press.

Du Bois, W. E. B. 2009. *The Souls of Black Folk*. B. H. Edwards (ed.). Oxford World's Classics. Oxford: Oxford University Press.

Du Bois, W. E. B. 2022. To the nations of the world. In J. Pitts and A. Getachew (eds), *W. E. B. Du Bois: International Thought*. Cambridge Texts in the History of Political Thought. Cambridge: Cambridge University Press, 18–21.

Fusco, K. and Olman, L.C. 2021. Techniques of justice: W. E. B. Du Bois's data portraits and the problem of visualizing the race. *MELUS* 46(3): 159–87.

Getachew, A. 2019. *Worldmaking after Empire: The Rise and Fall of Self-Determination*. Princeton, NJ: Princeton University Press.

Hartman, S. V. 1997. *Scenes of Subjection: Terror, Slavery, and Self-Making in Nineteenth-Century America*. New York: Oxford University Press.

Hartman, S. V. 2019. *Wayward Lives, Beautiful Experiments: Intimate Histories of Social Upheaval*. New York: W. W. Norton.

Murphy, B. J. 2021. 'Multiplied without number': Lynching, statistics, and visualization in Ida B. Wells, Mark Twain, and W. E. B. Du Bois. *American Literature* 93(2): 195–226.

Olman, L. C. 2022. Decolonizing the color-line: A topological analysis of W. E. B. Du Bois's infographics for the 1900 Paris Exposition. *Journal of Business and Technical Communication* 36(2): 127–64.

Ray, V. E., Randolph, A., Underhill, M. and Luke, D. 2017. Critical race theory, Afro-pessimism, and racial progress narratives. *Sociology of Race and Ethnicity* 3(2): 147–58.

Weheliye, A. G. 2015. Diagrammatics as physiognomy: W. E. B. Du Bois's graphic modernities. *CR: The New Centennial Review* 15(2): 23–58.

Wilson, M. O. 2012. *Negro Building: Black Americans in the World of Fairs and Museums*. Oakland, CA: University of California Press.

Wilson, M. O. 2018. The cartography of W. E. B. Du Bois's color line. In W. Battle-Baptiste and B/ Rusert (eds), *W. E. B. Du Bois's Data Portraits: Visualizing Black America*. New York: Princeton Architectural Press, 37–50.

Winters, J. R. 2016. *Hope Draped in Black: Race, Melancholy, and the Agony of Progress*. Durham, NC: Duke University Press.

Zuberi, T. 2001. *Thicker than Blood: How Racial Statistics Lie*. Minneapolis: University of Minnesota Press.

# 5

# A Feminist Ethic of Care for Orienting Utopia in Adjuntas, Puerto Rico

*Christie Nicoson*

**Introduction**

Everyday life in Puerto Rico (Borikén) exemplifies complicated politics of survival and hope in a changing climate. Networks of care feed neighbours, restore homes and share healthcare necessities amidst Caribbean storms, rising seas and successively hotter summers. These phenomena of climate impacts and mutual care occur amidst one of the oldest ongoing projects of colonisation in the world, starting with Spanish invasion in the 1500s and stretching to the United States of America's current occupation of Puerto Rico as an 'unincorporated territory'. Climate change here manifests physically but reveals colonial violences: US-imposed import-based food systems, military occupations and dependencies on fossil fuels influence how people experience climate events.

Puerto Rico and other Caribbean countries are often cast as highly climate vulnerable due to a combination of physical climate-related hazards (such as extreme hurricanes, rising sea levels and increasing surface and water temperatures) and unfavourable social-political conditions. Economies centred around tourism, agriculture and fisheries; largely import-reliant food and energy systems; and demographic legacies of racialised sterilisation and military interventions shaped by past and present colonialism all exacerbate and shape vulnerability (Jacobs et al. 2013; Sealey-Huggins 2017). Amidst seemingly endless depictions of future climate dystopias and insufficient climate action (see Foley, this volume), how might hopes of more utopian possibilities take shape? How does this hope inform politics and contribute to processes of transformation?

I grapple with these questions alongside the community group Casa Pueblo in Adjuntas, a town sitting amidst the mountains of Puerto Rico's largest island. Casa Pueblo first organised forty years ago as a movement to protect the people and environment against an open-pit mining proposal. Their eventual success in preventing the mining project prompted the group to ask their community, 'If not the mine, what instead?'. Alexis Massol González (2022, 42), one of the co-founders of the group, recalls asking: 'Is dreaming not allowed? Where is our sense of hope?' Based on fieldwork with Casa Pueblo conducted during 2022 and 2023, I study 'concrete utopia', a concept that rejects the idea of utopia as either an impossible, fantastical 'no place' or a blueprint with clear goals. Instead, this concept holds an aesthetic of an unknown together with active struggles to address injustices and work towards imagined possibilities. Casa Pueblo's ongoing struggles for social and environmental justice not only conjure questions and create space for hope, but also reveal 'educated hope' as a political tool for self-governance.

In this chapter, I follow José Esteban Muñoz's conceptualisation of concrete utopia in terms of 'educated hope' (2009). Muñoz uses utopia as hope, such that educated hope involves both imagination and grounded practices. A critique of the present and remembrance of the past educates political imagination and active struggle towards that which is 'not yet' here (2009). Thus, I study concrete utopias as manifestations of hope through political imaginaries and the construction of liveable alternatives. To explore not only what hope entails, but how it translates into everyday politics and implications of this for societies grappling with climate change impacts, I use a framework that takes a feminist ethic of care as an epistemic tool (Nicoson 2021). Caring values and practices both necessitate and produce knowledge (Puig de la Bellacasa 2017); employing an ethic of care approach turns attention to politics of knowledge in emergent processes of change. I study how this care shapes hope (imagined and in action) so as to analyse the orientation of utopias.

The chapter follows a narrative research design that is informed by a theoretical frame of a feminist ethic of care episteme. Following Muñoz's concrete utopia as educated hope (2009), I study instances of this hope enacted through *autogestión* in Adjuntas, Puerto Rico. I begin by discussing the context of Puerto Rico with examples of past and present visions of different futures. This not only establishes utopian dreaming and doing as ongoing processes, but also makes clear the need to better understand towards what or for whom utopian visions turn. Next, after presenting examples of concrete utopias in Adjuntas, I analyse these as processes through a lens of care-based knowledge. Finally, I use the concept of orientation to understand how this educated hope turns or shapes the community as a political project.

## Histories and Futures in Puerto Rico

In dominant historical accounts, habitation of the Puerto Rican archipelago begins with the Taino people, stretching back six or seven thousand years before first contact with Europeans. Considered one of the earliest Spanish colonies in the world, the occupation of Taino land stems from Christopher Columbus's 'discovery' of the Caribbean islands in 1492. In the early 1500s, Spanish invaders carried out genocide against the Indigenous peoples; they enslaved peoples kidnapped from Africa and around the Caribbean and built an economy around sugar plantations (Lloréns 2021). As part of the 1898 Treaty of Paris that ended the Spanish American War, Spain ceded Puerto Rico and other colonial territories to the United States. These early imperial infrastructures laid a foundation of enduring extractivism in Puerto Rico that has persisted through the institution of slavery, dominance of the sugar industry, and industrialisation via the 'economic development' programme Operation Bootstrap (Cabán 2002; Lloréns 2021; Santana 1998).

Throughout the occupation, the US government has led campaigns to crush resistance, promote extractive economies and entrench militarisation in the Puerto Rican archipelago. The government has targeted political resistance, including through the incarceration of nationalist and independence leaders (LeBrón 2019). Colonial financial infrastructure constructed an 'unpayable' debt of US$72 billion in bonds, plus US$50 billion in pension obligations, amassed through implementation of exploitative austerity instruments that benefit predatory investors while dismantling public services like healthcare and education, and entrenched by periods of economic recession (Morales 2019). The US military also has long occupied land, air and water; its bases and activities contaminate the environment and assault humans and more-than-humans through, for instance, military training exercises and chemical weapons testing (Cruz Soto 2008; Dickerson 2015; Sheller 2020). Moreover, targeted campaigns engender racial and gender disparities. For example, during the mid-1900s, local and federal governments implemented *la operación*, purporting poverty alleviation through tackling an 'overpopulation problem'. Laws in the 1930s encouraged sterilisation eugenics, enacted through mass sterilisation campaigns of *la operación* that targeted low-income women of colour for decades; by the 1980s, an estimated one third of Puerto Rican women were sterilised, ranking among the highest rates of sterilisation in the world (Womack 2020). These examples illustrate colonial dynamics that have shaped the islands' eco-social environment to this day.

Rather than passively submitting to this occupation and domination, Puerto Ricans have a long history of resistance and creation of alternatives. Amidst the colonial catastrophes to which impacts of climate change add, powerful utopian visions and revolutionary efforts persist. Early revolts and abolitionist movements succeeded in officially ending the institution of slavery in 1873; decades of struggle against the military in Vieques culminated in 2003, when protests led by women's movements succeeded in expelling the US Navy from the second largest island of Puerto Rico (Cruz Soto 2020). Anti-colonial struggles have connected with environmental justice movements, encompassing policies and practices to regulate contamination and degradation, resist mining projects and urban developments, and protect beaches and forests (Atiles-Osoria 2014; Santana 1993). Alternative visions of society take shape as community land trusts (Lloréns and Stanchich 2019), shared gardens and kitchens (Roberto 2019; Sou 2022) and brigades for clearing debris from streets or delivering care and recovery aid after hurricanes (Garriga-López 2019; Lloréns 2021; Unanue et al.).

On the other hand, there exist imaginaries of and efforts towards a more neoliberal or technocratic paradise. The so-called Jones Act, imposed in 1917, restricts transport to this day via stipulations that maritime cargo to and from Puerto Rico travel on ships under the US flag; this affects access to fuel, food and other basic goods and services (Torruella 2018; Valentin-Mari and Alameda-Lozada 2012). After the devastation of Hurricane Maria in 2017, this law prevented Puerto Rico from accepting offers of recovery assistance and supplies from neighbouring Dominican Republic and others. Instead, Puerto Ricans were forced to wait for relief from the United States, which proved overwhelmingly insufficient: supplies went to waste due to mismanagement and, for many, aid never materialised (ICADH 2017; Lloréns 2021).

More recently, Acts 20 and 22 (consolidated into Act 60 in 2020) allow outsiders from the US mainland and elsewhere to arrive, buy land and operate businesses in Puerto Rico while receiving tax breaks and other benefits (Klein 2018). These policies incentivise businesses and investors to use Puerto Rico as a base for cryptocurrency and blockchain operations, which consume massive amounts of electricity and appropriate property in ways that disrupt both political and economic power in Puerto Rico as a new form of settler colonialism (Crandall 2019). Local and federal neoliberal policies have gutted public services and infrastructure, including through the privatisation of the healthcare and energy systems in the 1990s and large-scale closure or privatisation of primary, secondary and higher education institutions in Puerto Rico (Brusi and Godreau 2019).

Imaginaries of a concrete utopia based on educated hope of community care and social justice stand in stark contrast to neoliberal visions of technological innovation and capitalist growth (see also Zebrowski in Chapter 2). The following section presents in greater depth a particular vision and enactment of a concrete utopia in Adjuntas, Puerto Rico, as a point from which to question and engage with political possibilities of educated hope.

## Concrete Utopia in Puerto Rico

Sociologist Ruth Levitas describes utopia as 'the expression of the desire for a better way of being or of living' (2013, xii). A concrete utopia, as introduced earlier, rejects the idea of impossible 'no-places' or pragmatic and carefully designed futures, instead preserving dreams of alternative possibilities in the tradition of Ernst Bloch's critical thought and Marxist philosophy (1986). Where abstract utopia lacks historic specificity, practical consciousness or political grounding, concrete utopia holds the aesthetic together with the political, the unknown of what will come with the struggle to address present and past injustices.

According to Muñoz, a concrete utopia as educated hope entails a (potential or actual) collective that engages with what it might become (Muñoz 2009, 3). Educated hope consists of both a critique of the present and a remembrance of the past – bringing forward the marginalising, exclusionary, oppressive or violent experiences and memories that are 'no longer conscious'. Critique and remembrance inform, or educate, political imagination and active struggle of that which is 'not yet' here, making possible affective visioning and struggle towards unknown alternatives (Muñoz 2009). The understanding of what is missing informs imagination about how a current situation could be changed. Such utopianism troubles linear thinking of time, moving beyond 'a preoccupation with liveness' (Warner 2011, 256). Beyond such constraints, concrete utopia entails both vision and method tied to the aesthetic, to beauty, to warmth as well as to social violence and the inevitable possibility of disappointment (Floyd 2010). Failure tempers such a utopian project, as reflexivity entails particular and partial values, desires or knowledge with gaps or missteps constrained by a wider context (Levitas 2013).

To study the transformations taking place in Adjuntas, I understand concrete utopia as educated hope (Muñoz 2009), as a performative process operating in the present that requires an envisioned, yet undetermined future together with struggles towards this. Described as 'performing impossibility in the face of the pragmatic' (Floyd 2010, 110), a concrete utopia is both imagined and grounded in concrete practices. In Puerto Rico the 'pragmatic' appears as an extension of colonial dystopia, one which suggests a future where rich outsiders get richer, and even a 'Puerto Rico without Puerto Ricans' (García López 2020). Meanwhile, a concrete utopia consists of people thriving, fulfilling needs for well-being in their specific place, as the case in Adjuntas illustrates.

In the 1980s, people in Adjuntas became concerned about plans for a large-scale open-pit mine proposed by the government and international corporations. Organising *Taller de Arte y Cultura*, they used local culture to resist and build support; instruments such as dance and theatre performances, *chiringas* (kites) festivals, murals, as well as tree plantings, demonstrations and petitions, bulletins, meetings and conferences demonstrated resistance and educated the community about potential social and ecological damage of the proposed mine (Massol González et al. 2008). Rallying around the call '*no a las minas, sí a la vida*' ('no to mines, yes to life'), support for the movement grew and in 1986 the government rejected the mining plan (Massol González et al. 2008).

After this initial success, the group turned to each other asking 'What next?' They concretised their movement, establishing the organisation Casa Pueblo and giving it a home through purchase and renovation of a house for the people. As Massol González says, they were 'sowing seeds of hope' (2022, 48) through *autogestión* centred around community voices: 'Having one's own voice as a self-governing community organisation allows us to set our own rules as we struggle and work towards reaching our goals' (Massol González 2022, 10). At Casa Pueblo, *autogestión* entails community self-governance through voluntary collaboration to address common interests and issues so as to improve their quality of life and foster a better future (Massol González 2022, 98). In addition, as Yarimar Bonilla (2020) notes, the term *autogestión* also describes 'developing a new being'; in the face of government and federal absence and failure, community-led conviviality undertakes necessary tasks of care as well as that of reimagining futures.

Asking 'what next', the land targeted for the mine stood at the forefront: if desires for the present and future did not include mining, what instead? *Autogestión* turned Casa Pueblo from *no a las minas* towards *bosque sí* (yes to the forest). They established Puerto Rico's first forest reserve managed by a self-governing organisation, *Bosque del Pueblo* (the People's Forest) (Massol González 2022). Casa Pueblo subsequently worked to establish solar energy for businesses, communal spaces and households – powering lights, cell phones, appliances and water pumps, while providing an alternative model for utilities, politics and a way of living. Reflecting on the project, current director of Casa Pueblo, Arturo Massol Deyá, charts the path from anti-mining to forest management and solar energy, noting how one activity led to another. As they move, they discover new tools to help achieve a yet-unknown vision:

> We learned that what we're doing is working on issues of water security; as we're protecting the forest, promoting an alternative economic model in which the primary people who will benefit from the activities are local people; [. . .] And as we stopped the mining and started with the forest conservation, we realised that the [fossil fuel-based] energy agenda was the greatest threat for the conservation of biodiversity. [. . .] And this

[solar energy] was a way for us to back up our ideas of moving away from fossil fuels, from the fossil fuel economy, with actions. (Massol Deyá interview 2022)

Casa Pueblo's experiences demonstrate imaginations of a future that remains undetermined; yet in 'walking the path', alternatives come into view. Action orients these imaginations towards hopes for something different, based in historicised reference to a common culture and place (Nicoson 2021). Here, *autogestión* takes the form of environmental work. Prevention of the open-pit mine protects natural resources in Adjuntas from extractivist exploitation, and subsequent preservation of the forest maintains the integrity of the environment for humans and more-than-humans locally and throughout the island of Puerto Rico:

We need to protect the natural resources and the integrity of the islands. So that's why we're defending the land from mining and natural gas. It's protecting because our nation is based on a common territory, Puerto Rico, the people, the culture and the setting. [. . .] Because Puerto Rico can't be in New York. We're Puerto Ricans in New York, because there's a Puerto Rico, an island. (Massol Deyá interview 2022)

Massol Deyá demonstrates that although the future is not specifically defined or planned, an aesthetic of place-based integrity is clear in this way of working towards a concrete utopia. The solar projects enable more equitable and reliable access to energy resources in Adjuntas, extending beyond basic provisioning or convenience. The cost of electricity in Puerto Rico is among the highest compared to the rest of the United States, coupled with some of the country's highest poverty levels (García Román, et al. 2020). The solar panels enable access to stable and affordable energy that remains available even when power plants go down or extreme weather events knock out power lines. Moreover, it maintains integrity of the inhabitants and place – allowing people to continue lives of dignity and well-being while also escaping the need for further development of fossil fuel-dependent energy systems.

**Orienting a Concrete Utopia**

How then is the hopeful *autogestión* in Adjuntas oriented? What does it do as part of a transformative process? Elsewhere, I have theorised that an ethic of care shapes how we know climate change, and that this way of knowing enables particular visions of the future in a process of climate transformation

(Nicoson 2021). A feminist ethic of care entails the values and practices for sustaining our world (Tronto 1993; 2013), which both produces and necessitates situated knowledge of contextual caring (Dalmiya 2016; Puig de la Bellacasa 2017). In this section, I take a feminist ethic of care as a focal point from which to orient a concrete utopia. By considering the way care orients and is oriented, we can begin to understand transformation distinct from current neoliberal visions present in Puerto Rico.

For queer theorist Sara Ahmed, orientation is 'about how we begin; how we proceed from "here"', because it 'expose[s] how life gets directed through the very requirement that we follow what is already given to us' (Ahmed 2006, 8, 21). Orientation shows how bodies inhabit space and how they turn, entailing elements of spatial and temporal background as well as repetition and use. Ahmed illustrates this through the metaphor of a writing table (2006): as a writer faces a table, sight focuses on the foreground, while other things fall behind the writer's back or into the background and out of their field of focus. Background also involves temporal dimensions, as past conditions affect what the writer perceives. Use ascribed to an object also orients: we know a writing table because we repeatedly use it to do the work of writing, as compared to a kitchen table used to prepare and serve food.

To use orientation so as to understand what utopias emerge and how they trouble linear or predetermined ideas about the future, I present the analytical tool outlined in Table 5.1. To each described phase of caring in the table (columns 1 and 2) (Tronto 2013), I pose questions about the orientation of care (column 3) in relation to an envisioned future (column 4). This tool enables consideration of concrete utopia as a metaphorical table: what is back- and foregrounded and for what and whose use. As such, I analyse how a care episteme orients Casa Pueblo's concrete utopia(s).

Key *autogestión* practices in Casa Pueblo include sales from Café Madre Isla coffee roastery and an artisanal shop that cover costs of operations; donations and maintenance of headquarters, which provide home and support for projects; and activities with the solar projects, *Bosque Escuela* (forest school) and *Bosque del Pueblo* that connect people and natural ecosystems in mutual caring. In addition to these, many more initiatives form the web of *autogestión*, including art and cultural events (choir concerts, murals, theatre performances, gallery exhibits and film screenings) and education programmes (at the *Bosque Escuela* and in science and music classrooms). Considering these through the tool in Table 5.1, caring values and practices originate and point to the community; a group of committed community members lead initiatives and inform the content of operations at Casa Pueblo, and people in Adjuntas are the primary targets and recipients of care. 'Experts' lend knowledge, skills

**Table 5.1** A feminist ethic of care perspective on orientations of utopia.

| Phase of caring | Description of phase | Guiding questions on orientation | Guiding questions on utopia |
|---|---|---|---|
| Caring about | Someone/thing identifies and assesses how to meet care needs | Who assesses care needs? Where/when does this assessment occur? Whose/what needs are in focus; whose are unaccounted for? | What histories and futures come into view? What falls from view? |
| Taking care of | Someone/thing takes responsibility for recognising and meeting needs | Who/what assumes this responsibility? | What is sustained? What is not sustained? How does (un)sustaining relate to envisioned futures? |
| Care-giving | Someone/thing directly meets needs | Who/what practises care? What use does giving care have? | How does a care-giver envision the future? What is the goal of the given care? |
| Care-receiving | Someone/thing responds as to whether needs are fulfilled by care received | Whose/what needs are fulfilled through caring? Whose/what care needs are/will be unfulfilled? | How does a care-receiver envision a future? What role do they play in an imagined future? |
| Caring with | Process of relations between care providers and receivers | Who/what are connected through the care processes? Who/what is outside this relation? | What caring is possible/not in existing relations? (How) are relations imagined otherwise? |

and labour to take care of and give care, for instance through installing solar panels on household and business rooftops or maintaining the ecological and educational systems and infrastructure at *Bosque Escuela*. These practices in turn create and foster relations between care-givers and care-receivers.

An image emerges of a body of people in Adjuntas with strong ties of solidarity and applied experience of care. The people lead one another towards continually realised and imagined futures of healthy eco-social environments and autonomous ways of living in relation with one another. New care needs emerge in terms of sustaining infrastructures – social networks, solar energy systems or production of consumer goods. This sustenance follows the self-described goal of *autogestión*, to be self-managing in the present and also to make way for something(s) new in the future. For instance, solar energy panels may prepare for future consumption needs and anticipate hurricanes that damage electricity grids. *Bosque del Pueblo* maintains the current ecosystem, while enshrining management of the land in the (present and future) local people, rather than leaving it open for projects of national development or private interests. *Bosque Escuela* engages people and ecosystems in the present so as to sustain this relation in the future: students learn about connections between societal and environmental systems, and nurture relationships that thrive in cycles of life, decay and regeneration that keep the ecosystem moving.

The ethic of care emphasises sustaining a healthy present and future for humans and more-than-humans. In addition, present and future relations are fostered in line with sustaining chains of care for an imagined autonomous future: strengthening local communal ties while simultaneously shifting relations of dependencies away from slow or absent state or federal government services. The communal ties and health in the present act in reciprocity towards this imagined future. For instance:

> People start to say, when we are putting up a solar panel, 'I think it is better they put it on that old lady's house because she has to undergo dialysis and needs energy.' (Massol González interview 2022)

In February 2023, a neighbourhood in Adjuntas collaborated with Casa Pueblo to install its own solar infrastructure of hopeful caring, describing it as transformation through self-managed 'solarization' of the community (Massol González 2022, 182; Zamira Padró 2023). With Casa Pueblo's facilitation, external donors provided funding and a local business installed solar panels to power public lighting and households for thirty-nine families. The project addresses burdensome costs of unreliable electricity; provides energy to maintain medications and healthcare for a community reporting a high occurrence of diabetes, asthma and high blood pressure; fosters everyday security through keeping streets lit; and prepares for cases of unknown events through the community-organised energy emergency fund (Zamira Padró 2023).

In an interview, Massol González describes how, after the success of protesting against open-pit mining, the Café Madre Isla project emerged hand in hand with plans to protect the forest. During community dialogues asking 'What next?', a group member pointed towards the hills, saying, 'the alternative is there' – and this suggestion inspired the coffee roastery that enables Casa Pueblo's independent funding (Massol González interview 2022). This orientation in turn opened new ways of being for individuals and for the wider community in Adjuntas, as seen in their solar energy projects. From situated (and thus also partial) histories and present experiences of energy shortages, people in Adjuntas recognise the need for electricity to secure medical support, store food, provide fresh water and enable livelihood activities. Massol González says:

> The government will not change the energy sources; they do not want to change away from oil and gas. So, the only way to make change is from the bottom up, to pressure them. This organising is a form of resistance, and also a form of an alternative. (Massol González interview 2022)

He demonstrates how a care episteme orients concrete utopia: the well-informed mining protest oriented them towards the forest, a turn which made visible threats posed to the environment by the continued dependence on fossil fuels as well as opportunities for independent financing through Café Madre Isla. A dialect of care between givers and receivers orients Casa Pueblo to address people's energy needs alongside environmental needs, pointing them towards solar projects. Returning to Ahmed's metaphor (2006), I suggest that what once stood as a board with legs (a generic table) was increasingly used and perceived otherwise (turning it from 'any' table to a desk): protesting against the mine increasingly involved a movement to protect the people of Adjuntas. This evolved through protecting the surrounding environment from degradation, ensuring the sustenance of a place-based identity and securing basic needs.

## Conclusion

In this case from Adjuntas, the ethic of care episteme demonstrates that complexity matters and dependencies have been reoriented. For instance, rather than relying on privatised electricity infrastructure and government response, the community built a new means of energy production. As for different phases of caring, Casa Pueblo conducts 'caring about' and 'taking care of' through dialogue processes, and they share in care-giving and care-receiving with members of the public. In addition to individual involvement, private companies and foundations also take part in care-giving, through provision of essential funds and equipment to construct energy infrastructure. Although projects at Casa Pueblo still entail dependencies, and also crucially rely on non-local actors, the recognition of needs and responses to fulfil those needs are rooted in the social and environmental community of Adjuntas. The phases of caring stem from and orient towards relations between land and people – those in Adjuntas, throughout the islands and in the Puerto Rican diaspora.

With this chapter, I do not strive to uncover a perfect blueprint for climate utopias. Rather, I share insights from Casa Pueblo's ongoing processes to illustrate the complexities of social transformation and projects of concrete utopias. The discussion here provides examples of self-governance in climate transformation and shares stories of everyday lives and politics in changing climate landscapes under ongoing coloniality. These examples demonstrate the political power of educated hope for disrupting a status quo and opening alternative ways of being. Ultimately, this contributes to contesting ideas about islands doomed to dystopian climate futures by theorising ways of engaging with hope in concrete utopias and politics for a future that need not be in a distant temporal timeline.

## References

Ahmed, S. 2006. *Queer Phenomenology: Orientations, Objects, Others*. Durham and London: Duke University Press.

Atiles-Osoria, J. 2014. Environmental colonialism, criminalization and resistance: Puerto Rican mobilizations for environmental justice in the 21st century. *RCCS Annual Review* 6.

Bloch, E. 1986. *The Principle of Hope*. Oxford: Blackwell.

Bonilla, Y. 2020. The coloniality of disaster: Race, empire, and the temporal logics of emergency in Puerto Rico, USA. *Political Geography* 78: 102181.

Brusi, R. and Godreau, R. 2019. Dismantling public education in Puerto Rico. In Y. Bonill and M. LeBrón (eds), *Aftershocks of Disaster: Puerto Rico Before and After the Storm*. Chicago, IL: Haymarket Books, 234–49.

Cabán, P. A. 2002. Puerto Rico: State formation in a colonial context. *Caribbean Studies* 30(2): 170–215.

Crandall, J. 2019. Blockchains and the 'chains of empire': Contextualizing blockchain, ctryptocurrency, and neoliberalism in Puerto Rico. *Design and Culture* 11(3): 279–300.

Cruz Soto, M. 2008. Inhabiting Isla Nena: imperial dramas, gendered geographical imaginings, and Vieques, Puerto Rico. *CENTRO Journal* xx(1): 165–91.

Cruz Soto, M. 2020. The making of Viequenses: Militarized colonialism and reproductive rights. *Meridians* 19(2): 360–82.

Dalmiya, V. 2016. *Caring to Know: Comparative Care Ethics, Feminist Epistemology, and the Mahābhārata*. New Delhi: Oxford University Press.

Dickerson, C. 2015. Secret World War II chemical experiments tested troops by race. *NPR Morning Edition*. Available at https://www.npr.org/2015/06/22/415194765/u-s-troops-tested-by-race-in-secret-world-war-ii-chemical-experiments (accessed 2 January 2024).

Floyd, K. 2010. Queer principles of hope. *Mediations* 25(1): 107–13.

García López, G. A. 2020. Environmental justice movements in Puerto Rico: Life-and-death struggles and decolonizing horizons. *Society & Space*. Available at https://www.societyandspace.org/articles/environmental-justice-movements-in-puerto-rico-life-and-death-struggles-and-decolonizing-horizons (accessed 2 January 2024).

García Román, A. G., Rivera, A. I., Jaramillo, A. F., Deyá A. M., Kunkel, K., Ortiz, D., Carrillo O'neill E., Vila Biaggi, I. M., Menéndez, J., Rosario, J., Rodríguez Rivera, L. E., Conty, M., Saadé, P., Santiago, R. and de Castro, V. 2020. *Queremos Sol: Sostenible, local, limpio*. Queremos Sol PR.

Garriga-López, A. 2019. Puerto Rico: The future in question. *Shima: The International Journal of Research into Island Cultures* 13(2): 174–92.

ICADH. 2017. *Justicia Ambiental, Desigualdad y Pobreza en Puerto Rico*. Facultad de Derecho de la Universidad Interamericana; Instituto Caribeño de Derechos Humanos.

Jacobs, K. R., Ernesto L. D., Carrubba, L., Castañer, J. A., Chaparro, R., Crespo Acevedo, W. L., Diaz, E., Espinoza, R., Gaztambide, S., Kasey R. Jacobs, Moyano, R., Nieto, V., Pacheco, S., Ramos, A., Santa, P., Santiago Bartolomei, R., Santini Rivera, R., Seguinot Barbosa, J. and Terrasa, J. J. 2013. Working Group 3 Report: Climate change and Puerto Rico's society and economy. In *Puerto Rico's State of the Climate 2010–2013: Assessing Puerto Rico's Social-Ecological Vulnerabilities in a Changing Climate*. San Juan: Puerto Rico Climate Change Council.

Klein, N. 2018. *The Battle for Paradise: Puerto Rico Takes on the Disaster Capitalists*. Chicago, IL: Haymarket Books.

LeBrón, M. 2019. Building accountability and secure future: An interview with Mari Mari Narváez. In Y. Bonilla and M. LeBrón (eds), *Aftershocks of Disaster: Puerto Rico Before and After the Storm*. Chicago, IL: Haymarket Books, 319–31.

Levitas, R. 2013. *Utopia as Method: The Imaginary Reconstruction of Society*. Basingstoke and New York: Palgrave Macmillan.

Lloréns, H. 2021. *Making Livable Worlds: Afro-Puerto Rican Women Building Environmental Justice*. Seattle: University of Washington Press.

Lloréns, H. and Stanchich, M. 2019. Water is life, but the colony is a necropolis: Environmental terrains of struggle in Puerto Rico. *Cultural Dynamics* 31(1–2): 81–101.

Massol Deyá, A. 2022. Interview. Conducted by Christie Nicoson in Adjuntas, Puerto Rico, 14 July 2022.

Massol González, A. 2022. Interview. Conducted by Christie Nicoson in Adjuntas, Puerto Rico, 8 July 2022.

Massol González, A. 2022. *Casa Pueblo: A Puerto Rican Model of Self-Governance*. A. Ravikumar and P. A. Schroeder Rodríguez (trans.). Lever Press.

Massol González, A., Johnnidis, A. A. and Massol Deyá, A. 2008. The evolution of Casa Pueblo, Puerto Rico: From mining opposition to community revolution. *Gatekeeper*. IIED.

Morales, E. 2019. Puerto Rico's unjust debt. In Y. Bonilla and M. LeBrón (eds), *Aftershocks of Disaster: Puerto Rico Before and After the Storm*. Chicago, IL: Haymarket Books, 211–23.

Muñoz, J. E. 2009. *Cruising Utopia: The Then and There of Queer Futurity*. New York and London: New York University Press.

Nicoson, C. 2021. Feminist peace and the politics of climate change: Ethic of care orientations toward climate transformation. EISA Pan-European Conference on International Relations [virtual], 14 September 2021.

Puig de la Bellacasa, M. 2017. *Matters of Care: Speculative Ethics in More Than Human Worlds*. Minneapolis: University of Minnesota Press.

Roberto, G. 2019. Community kitchens: An emerging movement? In Y. Bonilla and M. LeBrón (eds), *Aftershocks of Disaster: Puerto Rico Before and After the Storm*. Chicago, IL: Haymarket Books, 309–18.

Santana, D. B. 1993. Colonialism, resistance, & the search for alternatives: The environmental movement in Puerto Rico. *Race, Poverty & the Environment* 4(3): 3–5.

Santana, D. B. 1998. Puerto Rico's Operation Bootstrap: colonial roots of a persistent model for 'Third World' development. *Revista Geográfica* 124: 87–116.

Sealey-Huggins, L. 2017. '1.5 °C to stay alive': Climate change, imperialism and justice for the Caribbean. *Third World Quarterly* 38(11): 2444–63.

Sheller, M. 2020. *Island Futures: Caribbean Survival in the Anthropocene*. Durham, NC and London: Duke University Press.

Sou, G. 2022. Reframing resilience as resistance: Situating disaster recovery within colonialism. *The Geographical Journal* 188(1): 14–27.

Torruella, J. R. 2018. Why Puerto Rico does not need further experimentation with its future: A reply to the notion of 'territorial federalism'. *Harvard Law Review Forum* 131(3): 65–104.

Tronto, J. 1993. *Moral Boundaries: A Political Argument for an Ethic of Care*. London and New York: Routledge.

Tronto, J. 2013. *Caring Democracy: Markets, Equality, and Justice*. New York: New York University Press.

Unanue, I., Patel, I., Tormala, T. T., Piazza Rodríguez, A. A., Méndez Serrano, K. and Brown, M. 2020. Seeing more clearly: Communities transforming towards justice in post-hurricane Puerto Rico. *Community Psychology in Global Perspective* 6(2): 22–47.

Valentin-Mari, J. and Alameda-Lozada, J. I. 2012. *Economic Impact of Jones Act on Puerto Rico's Economy*. Working paper presented to US General Accountability Office.

Warner, S. 2011. Review 'Cruising Utopia: The Then and There of Queer Futurity'. *Modern Drama* 54(2): 255–7.

Womack, M. L. 2020. US colonialism in Puerto Rico: Why intersectionality must be addressed in reproductive rights. *St Antony's International Review* 16(1): 74–85.

Zamira Padró, N. 2023. Comunidad Alto de Cuba comienza su transformación a un sistema energético solar. *NOTICEL*. Available at https://www.noticel.com/vida/top-stories/20230220/comunidad-alto-de-cuba-comienza-su-transformacion-a-un-sistema-energetico-solar/ (accessed 2 January 2024).

# Part II
# Governance

# 6

# Enduring Hopelessness: Governance without Horizon in Pandemic Times

*Nicolas Gäckle*

**Introduction**

Exhaustion and fatigue have gradually become markers of pandemic life across various discursive and practical fields. Medical discourse describes fatigue as 'one of the most persistent and debilitating post-Covid-19 symptoms' (Azzolino and Cesari 2022; see also NHS England 2021). The psy-sciences call for (self-)'managing lockdown fatigue' (Australian Psychological Society 2020), urging subjects to practise healthy ways of living through COVID's suspension of normality. Critical interventions suggest that the pandemic has amplified a 'collective fatigue' (Han 2021; see also Gemignani and Hernández-Albújar 2022) originating in neoliberal modes of governance. The importance of these connections are confirmed by the World Health Organization (WHO) which has linked pandemic fatigue to 'complacency, alienation and hopelessness' (WHO 2020c, 7). The organisation has tried to gain an understanding of the phenomenon through a series of reports, tools and statements (WHO 2020c, 2020d, 2020e, 2021; see also Gäckle 2023).

Departing from the broad appeal that the proliferation of the semantic of fatigue indicates, this chapter explores how the problematic link between fatigue, hopelessness and horizonlessness enters the governmental imaginary underlying the COVID-19 pandemic. Fatigue is understood as an emergent affective correlate of the encounter with 'viral time' (Flexer 2020). Responding to this viral temporal regime, characterised by the looming potential and occasional actualisation of mutation, pandemic governance opts for the continuous protraction of the emergency present (see Grove et al. 2022; Rabi et al. 2022). Whereas emergency governance traditionally operates within an interval of hope and the possibility of the alterability of the future, and hence of overcoming the cut in time manifested by the crisis (Anderson 2017), the problematisation of fatigue grasps onto a hope of keeping this cut in temporal suspension.

Instead of overcoming crisis and striving for a better future, the problematisation of fatigue emphasises the need to endure the present. Fatigue becomes a marker of what Pamela Lee has aptly termed the 'grim phenomenology of the present tense' (2021, 4) that problematises liberal imaginaries of governance by way of its very horizonlessness. Its appeal within the WHO signals a parallel movement of governmentalisation that thematises the absence of aspiration and seeks to capture the 'sense of stuckness' of a protracted emergency (Honig 2009, 10; see also Adey et al. 2015, 9). It becomes one of the 'new descriptors for understanding the temporality' of an 'extended "emergency present"' (Anderson 2017, 472). This engagement with the problems associated with an experience of temporal suspension subsequently renders the pandemic as an experimental space for rehearsing the coming crises of the Anthropocene (Latour 2021).

Empirically, this chapter explores the WHO's problematisation of pandemic fatigue in order to contribute to previous work on the governmentalisation of hope (Anderson 2014; Stäheli 2014; Wrangel 2019). It further links the problem of the loss of affect for hope to recent debates on emergency response (Anderson 2017; Anderson et al. 2020; Zebrowski 2019), emphasising fatigue's blurring of everyday/emergency distinctions. Lastly, it reiterates the centrality of the pandemic rupture of inherited notions of time (Fassin and Fourcade 2021; Kattago 2021; Ruwet 2021) by following the attempt to chronopolitically organise endurance within the durational terrain of a horizonless present (Gorfinkel 2012; Lisle 2023; Reid 2023).

To pursue this exploration, the first section situates the problematisation of pandemic fatigue within a reading of emergency governance that highlights how its presentist orientation responds to the specific temporality proper to COVID-19. Subsequently, the chapter sketches fatigue's problematic relation to the horizon as a cornerstone of imaginaries of liberal governance. The last section transfers these considerations to the empirical site of pandemic fatigue. It highlights how pandemic governance deals with COVID's temporal suspension through an attempt to cultivate endurance.

**Finding Hope in Emergency**

Under the auspices of a state of exception paradigm, emergencies have long been read as opportunities for the consolidation of governmental power (Anderson 2021b, 16; see also the review of the literature in Zebrowski in Chapter 2). Most prominently, Giorgio Agamben (1998; 2005) has popularised such a perspective through the observation of a paradox: the sovereign ability to legally operate outside of law. Sovereignty and biopolitics merge in the state's capacity to decide which lives to value and which to abandon as bare life. Following Agamben, the 'voluntary creation of a permanent state of emergency [. . .] has become one of the essential practices of contemporary states' (2005, 2), unsurprisingly also operating in the governance of the COVID-19 pandemic (Agamben 2021).

The state of exception is, however, only one specific way of governing emergencies (Adey et al. 2015, 4). In the context of the COVID-19 pandemic, Sergei Prozorov (2021a) and Benjamin Bratton (2021) have emphasised that Agamben's hyperbolic timbre not only occludes the specific rationalities developed to confront the pandemic but outrightly ignores the material, biological and experiential reality of the pandemic. The stubborn insistence on the validity of the state of exception paradigm undermines a critical project attuned to how governance '*takes form*' in relation to emergencies' (Anderson 2021b, 1360, italics in original). While the state of exception imagines an omnipotent, malicious state actor, starting from the phenomenon which is declared an emergency itself may indicate the aspects of reality exceeding governmental capture and thereby motivating governmental innovation.

The temporal emergence of COVID-19 is one crucial excessive feature against which pandemic emergency measures take form. 'Viral time' (Flexer 2020) emerges from a complex 'biosociological' (Foucault 2003, 250) amalgam. As such, it integrates both the different temporal properties that the virus produces – such as incubation and contagion periods, the course of infection or the capacity of the virus to survive on different surfaces – and the 'existing social and productive relations' within which those temporalities unfold and produce differential spread dynamics (Flexer 2020, 4; see also Weber 2022). Where such conditions prove favourable, exponential contagion may be the result, adding an axis of acceleration to the temporal picture (Ruwet 2021). The early months of the pandemic saw knowledge about these temporal cornerstones emerging from scientific studies. A highly dynamic 'ground-zero empiricism' gradually tamed the 'radical uncertainty' instantiated by the pandemic (Daston 2021, S56; Ruwet 2021).

However, given the capability of viruses to become different by 'reinventing themselves by mutation and adaptation' (Voelkner 2019, 382) – particularly where infections increase – the horizon of expectation outlined by such scientific findings is necessarily provisional. This epistemic preliminarity is a direct effect of the emergent viral ontology. As the sciences mobilised to examine the virus stress its complex and contingent becoming, they inscribe emergence onto the imaginary of life as the biopolitical object of governance (Dillon 2007; Dillon and

Lobo-Guerrero 2009). The generalised public experience of how this epistemic precarity translates into potentially fast-changing policies is one of the striking novelties of COVID-19.

Strategies for managing the pandemic hence took into account the emergent qualities of viruses as 'contingently performative as well as situated, and coevolving with their environments' and thus 'capable of moving out of phase with themselves and becoming other than what they were' (Dillon 2007, 14). While a range of innovative security technologies such as preparedness (Lakoff 2007) and scenario planning (Krasmann and Hentschel 2019; Samimian-Darash 2022), as well as pre-emption (Cooper 2006; Roberts and Elbe 2017) and pre-mediation (Grusin 2010), have been analysed as attending to the problematique of emergence, they unanimously focus on rendering the future governable. The COVID-19 pandemic, however, poses the question of emergence not on the register of a looming potential but of a present actualisation (Rabi et al. 2022; Ruwet 2021).

Accordingly, emergency response techniques, engaging with the curious viral temporal markers of the pandemic, affirmed the presentist outlook that 'events cannot be wholly predicted and prevented' (Zebrowski 2019, 149) and instead need to be dealt with 'as and when [they] happen' (Anderson 2016, 15). Given that infections could not be completely avoided, calls to 'flatten the curve' have epitomised a tactic that intervenes on an emergent dynamic through stalling infections and thus attempting to prevent hospitals from reaching capacity limits.[1] Pandemic emergency response thus addressed the non-linear spread of the infectious disease through a chronopolitical tactic of protraction and slowing down (Ruwet 2021). It sought to 'strip an emergent event of its disruptive potential' mainly by engaging in a longer-term effort of social distancing and other protective measures (Zebrowski 2019, 149; see also Ruwet 2021). Aimed at temporal suspension rather than a quick return to normality, this tactic destabilised the intervallic character of emergency and its relation to hope.

Ben Anderson understands the interval of emergency as 'the gap or break during which emergency action can still make a difference' (2017, 470), provided it is timely and decisive. On the one hand, the interval produces a *kairotic* 'transformational time of action' (Hutchings 2008, 5; see also Smith 1969) and a sense of urgency. On the other hand, it is tied to a promise of overcoming that is squarely linked to hope. The specific orientation of hope depends on how it relates to the 'normality' of the everyday.

From the governmental logic of emergency response, hope is constituted as a will to overcome the 'momentary, arrhythmic disruptions to the banal ebb and flow of everyday routine' (Grove et al. 2022, 9). It hence affirms a cyclical imaginary of time that yearns for the return to a normality that is seldom questioned. Emergency response maintains the value of regularity as a hallmark of healthy liberal lives. Following an observation by Henri Lefebvre, '[r]hythms unite with one another in the state of health, in normal (which is to say normed!) everydayness' whereas arrhythmia is 'symptom, cause and effect' of a discordance that 'brings previously eurhythmic organisations towards fatal disorder' (Lefebvre 2013, 25). Resonating with the genealogy of crisis governance in premodern medicine, *kairos* (the felicitous moment of intervention) is inherent to the specific condition that demands a 'cure' (Prozorov 2021b, 439). In the context of the COVID-19 pandemic, the conservative hope of emergency response thus requires modelling governmental action adaptively on the temporal emergence of the pandemic. However, in its stretched chronicity, the pandemic event tends to exceed a semantic so focused on the momentary.

Contrasting with the conservative notion of hope and its figuration of overcoming as return, other discourses have highlighted how endemic forms of 'slow violence', such as anti-Black violence or environmental degradation, go unnoticed within the eventful imaginary of emergency (Nixon 2011; see also Anderson 2017; Anderson et al. 2020). The difference is key: whereas governmental emergency response focuses on overcoming the *kairotic* cut in time and on pursuing a timely and decisive response (that assumes an almost natural link with affects of urgency), slow emergencies assume a durative, stretched temporality that blurs the distinction between emergency and everyday. Uncovering the audacity of a hope invested in a return to normality that only ever existed for some privileged portions of the population (Grove et al. 2022), hope here is more akin to a Blochian 'not yet'. Committed 'to change rather than repetition' (Bloch, in Anderson 2017, 474), it strives to affirm the openness of the future and thus to go beyond the mere return to normality. As such, it reorients hope towards the present in providing 'a dynamic imperative *to action* in that it enables bodies to *go on*' (Anderson 2006, 744; see also Schlegel 2022).

To some extent, the COVID-19 pandemic blurs the straightforward juxtaposition between a governmental insistence on (and an activist challenge to) the emergency/everyday distinction. A *Nature Human Behaviour* article authored by WHO experts and behavioural scientists and later taken up in the WHO report (2020c) on pandemic fatigue may be used for illustration. It articulates the need to produce public awareness that 'there will be no going back to normal but a stepwise approach to a "new normal"' and explicitly curbs hopes for 'an immediate return to normal' (Habersaat et al. 2020, 678). As a premature desire for a return to normality, hope itself becomes a problem during the pandemic. What is needed instead is a 'phased approach to a new normal' that evolves adaptively with the epidemiological situation, involving 'adjustments to restrictions and potential re-employment of previous stricter measures' (Habersaat et al. 2020, 678). Required to mirror the wave-like emergence of infections, pandemic response turns the intervallic into a condition, recalibrating its activity towards a chronic alternation between phases of more and less severe restrictions. The article further contrasts the *kairotic* imaginary of emergency with the need to slowly adopt new routines and rhythms of the everyday (Habersaat et al. 2020, 679). These examples suggest a reformulation of the governmental imaginary of emergency that stresses its pandemic everydayness (see Grove et al. 2022, 9).

The COVID-19 pandemic confounds the expectation that momentary urgent action will help in quickly overcoming the emergency. Instead, it calls for endurant modes of settling into the exception. Pandemic governance is thus less concerned with formulating different futural temporal aspirations – a conservative hope for the return to 'normality' or a progressive hope for its overcoming – but with finding ways to deal with the absence of such a temporal horizon. This absence requires a reconsideration of questions of hope and hopelessness which the next section discusses by looking at how the notion of fatigue relates to the pandemic lack of the horizon as an element structuring liberal imaginaries of governance.

**The Horizonlessness of Fatigue**

The notion of overcoming is deeply linked to the function of the horizon as a metaphor structuring liberal imaginaries of governance and subjectivity. Underwriting the promise of a progressive understanding of historical time (Koselleck 2004), the horizon serves as a temporal background against which violence in the present is disavowed with reference to future improvement (Povinelli 2021, 38; see also Chandler in Chapter 1). Where the horizon enjoys an appealing plausibility, 'the present is more or less a problem to be solved by hope's temporal projection' (Berlant 2011, 13), even if the attachment to this unredeemed promise is cruel. Hope is appointed to a future and affectively present – 'even expectation can be experienced' (Koselleck 2004, 261) – for aspiring liberal subjectivities (Baraister 2017, 8; Lobo-Guerrero 2016, 27; Reid 2023, 12).

Hope is tied to an experience in which time unfolds towards a horizon in progressive motion (Baraister 2017, 6). Premised on the ability to appropriate moments of crisis as moments of reassertion, this horizon holds the promise of a liberal meta-stability (Lobo-Guerrero 2016, 28; see also Bargués-Pedreny 2019, 279). While a once-appointed hope may be dis-appointed as expectations go unfulfilled, their non-fulfilment only ever fuels the desire to anticipate more accurately next time.[2] Consequently, however, situations in which time's motion is experienced as suspended exceed these premises. They require a grammar of stuckness, non-movement, vectorlessness, or, in other words, a form of horizonlessness that poses the problem of hope not in terms of dis-appointment but as non-appointment. Lauren Berlant (2011), Elizabeth Povinelli (2011; 2021) and Lisa Baraister (2017) have interrogated modes of eventfulness beyond the progressive and the exceptional, making visible the unspectacular 'crisis ordinariness' (Berlant 2011, 10) characteristic of the experience of chronic violence. It is precisely the absence of the horizon and its implied experience of linear time that may produce exhausting effects.

Inhabiting 'a zone of temporality marked by ongoingness, getting by, and living on' (Berlant 2011, 99) thus extorts endurance rather than the resilience of subjects (see Salazar and Scheerder 2022). Whilst bouncing back or progressively overcoming have become hallmarks of the affinity between resilience and neoliberalism, they remain tied to a horizon that is hoped for. Endurance instead articulates a mode of durational, horizonless and hopeless being-in-the-world. While resilience reasserts subjectivity vis-à-vis the unfolding of time, endurance feeds on submission to it and may tip over into fatigue. Enduring means persevering in the absence of temporal markers. Fatigue finds endurance overburdened. It is 'inexorably a concern of and for duration' encompassing both, the affective presence of horizonlessness and the implication that there may be something beyond fatigue which demands its liminal passage (Gorfinkel 2012, 314f.).

Read through Bruno Latour's (2021) comment that the COVID-19 pandemic constitutes an Anthropocenic dress rehearsal, fatigue anticipates the immediate and generalised experienceability of the non-defensibility of modern divides between humans and nonhumans (see Hartog 2022, 228). As viral time emerges through entanglements and frequently exceeds control, it poses the problem of 'a time that is no longer ours' (Latour 2021, S26). The World Health Organization's concern for pandemic fatigue can be read as a governmental response that engages with the apparent subjective sense of viral time's horizonlessness. Modes of critical theorising have called attention to how 'processes that have not yet found their genre of event' (Berlant 2011, 4) lack recognition within predominating governmental imaginaries. However, we need to entertain the idea that with the pandemic, the experience of time 'as radically immoveable' and 'radically suspended' (Baraister 2017, 4) already finds itself in the process of being ontopolitically affirmed, gauged and appropriated within emergent modes of governance (Chandler 2018).

Despite such attempts, notions of fatigue as horizonlessness (and thus hopelessness) point to a figuration of subjectivity that is difficult to accommodate within liberal imaginaries of governance. While dis-appointed hope may still motivate aspiring subjects to bounce back and overcome, hopeless subjects – unwilling or unable to appoint a future – are potentially much more troubling. As a form of exhaustion, hopelessness questions the very possibility of the horizon as a heretofore unquestioned background against which liberal lives acquire meaning: 'the exhausted exhausts all of the possible' (Deleuze 1995, 3). Tentatively, the task following here is one of reorienting governmental imaginaries in light of a durational present characterised by its horizonlessness, in which the subject must endure whilst avoiding a tipping over into fatigue.

Having discussed the horizonlessness of fatigue, the next section traces its entrance into pandemic imaginaries of governance and subjectivity through the problematisation of 'pandemic fatigue'.

## Tired of Hoping

The WHO Regional Office for Europe engaged with pandemic fatigue for the first time in autumn 2020, summarising the results of a series of behavioural insights studies and outlining possible responses to decreasing adherence to protective measures, in the report *Pandemic fatigue: Reinvigorating the Public to Prevent COVID-19* (WHO 2020c). Picking up on the preceding considerations, this section illustrates the horizonlessness of fatigue. It reconstructs how the notion creates friction with the idea of a quick return to normality and how it proposes a model of rhythmical living in the register of subjectivity.

### *Problematising Horizonlessness*

Pandemic fatigue is deemed problematic as a form of slow public detachment from the ongoing pandemic emergency. Its immediately observable surface – behaviours increasingly contradicting public health recommendations – is complemented by an affective layer. Fatigue expresses a sense of weariness that reflects the stretched temporality of the pandemic through the gradual emergence of 'complacency, alienation and hopelessness' as well as 'stress, loneliness and boredom' (WHO 2020c, 7, 19). It further alienates subjects by challenging their sense of self-efficacy (WHO 2020c, 8).

By pointing to the gradual experiential normalisation of 'outrageous circumstances' (WHO 2020c, 8), pandemic fatigue rubs up against an imaginary of emergency whose activating character is to be channelled to precipitate a quick restoration of everyday normality. It biopolitically problematises the population as a 'global mass' inert in the face of emergency (Foucault 2003, 242). As such, articulations of fatigue are less concerned with forms of *active* resistance to emergency measures – protests are only marginally considered

(Habersaat et al. 2020, 678; WHO 2020a, 7) – but rather with unspectacular, passive registrations of a loss of affect. Fatigue hence insinuates that the horizon of normality is deceptive. As people aspire to this horizon, thinking 'that they are or soon will be returning to normal, their actions may hasten the onset of a second wave of the outbreak' (Habersaat et al. 2020, 678 f), they perform a proto-cruel-optimist attachment. Striving for normality thus becomes another factor in the prolongation of its suspension, a desire standing in the way of its own fulfilment (Berlant 2011, 1). The World Health Organization's problematisation of pandemic fatigue thus signals a governmentalisation of the affective implications of living through a protracted emergency. This is based, in turn, on an ontopolitical affirmation of the significance of the (subjective) experience of time for the emergence of the pandemic situation (Gäckle 2023).

Beyond complicating notions of returning to normality, pandemic fatigue also qualifies the anaesthetic logic of emergency response (Zebrowski 2019, 160; see also Zebrowski in Chapter 2). As the bulk of protective measures is aimed at instilling an everydayness of emergency that operates via a change of habits (see Habersaat et al. 2020), the pandemic stretches the temporal terrain of emergency response. While quelling security affects to close down chaos is necessary in circumstances where emergency materialises in a momentary manner, the problematisation of pandemic fatigue emphasises that the protraction of emergency measures itself has anaesthetic effects that need to be countered. Just as much as pandemic fatigue is understood to be a reaction to the temporal emergence of the pandemic itself, it is seen to be responding to the governance of this emergence. To counter fatigue, governmental communication needs to balance 'both under- and over-cautiousness among the public' (Habersaat et al. 2020, 680) whilst strengthening feelings of self-efficacy through positive, future-oriented wording such as 'progress, advance, community, cohesion, improve, perspective, reasonable, resourceful, optimistic and generous' (Habersaat et al. 2020, 680). Protective behaviour needs to be seen not as 'a matter of capitulation to authority or a reflection of despair, but a part of something positive, hopeful and (if at all possible) fun' (WHO 2020c, 13).

Countering fatigue through hope builds on a reading of subjectivity through self-efficacy as articulated by research in health communication and psychology. The value of hope here lies in its capacity to bring about behavioural change, or, in the case of the pandemic, to ensure endurance:

> the cognitions associated with self-efficacy (i.e., confidence in one's ability to perform an action) associate with the cognitive appraisals that generate hope. Relatedly, feelings of hope serve as a motivator to act, thus increasing the likelihood that self-efficacy will translate into actual behavior. (Nabi and Myrick 2019, 465)

Following Claes Wrangel, hope thus serves as a biopolitical technology that does not follow the aim 'to modulate the ontology of the world, but how this ontology is felt, perceived and acted upon' (Wrangel 2019, 670; see also Wrangel in Chapter 7). Employed as a mere functional motivator of feelings of self-efficacy, hope in this context is devoid of futurity. Instead, its affective appeal is strategised for a continuous preservation of endurance. Seen through an analytics of self-efficacy the endurant pandemic subject maintains a belief in the possibility of actively managing pandemic entanglements whilst avoiding exuberant aspirations. As a counter figure, the fatigued subject loses either confidence in its capability to act in an efficacious manner or the willingness to do so.

## Rendering Subjectivity Rhythmical

Pandemic fatigue signals a lack that people perceive in their subjectivity. In 'a constantly changing and restricting environment' their 'urge for self-control and self-determination' (WHO 2020c, 7) cannot be lived out. With respect to the economic field Urs Stäheli has argued that hope is employed to create 'future-oriented modes of subjectivation' (Stäheli 2014, 283, author's translation). In contrast, pandemic governance gauges the possibility of a self-efficacious hope without futurity, as reflected in a recurrent emphasis on the merits of a rhythmised everyday and modes of filling time (Flexer 2020, 5). This hope is focused on recurrence and repetition rather than progression. In the absence of futurity, presentist rhythmicity takes its place.

The WHO, for instance, sketches the need to '[c]reate opportunities for people to fill their time productively if isolated or unemployed due to the pandemic', followed up by the encouragement to '[h]elp them build more structure into everyday life and engage in activities that have positive effects, such as being outdoors, exercising and enjoying safe socializing' (WHO 2020c, 20). In the absence of futural markers, it becomes imperative to produce 'something that resembles normal life' in the here and now (WHO 2020c, 16). Rhythm and repetition counter fatigue's inability to anticipate. Similarly, the discomfort and stress caused by the exposure to perpetual news updates – linked to pandemic fatigue through the decreasing motivation to stay informed about the pandemic (WHO 2020c, 7) – may be countered by designating 'specific times during the day' for gathering this information (WHO 2020b, 5). On a broader scale, successfully normalising 'new behaviours in old environments [. . .] is a function of repetition' (Habersaat et al. 2020, 679). Introducing those moments of rhythmic micro-structuring is therefore embedded in the larger-scale learning of new routines.

Managing the pandemic emergency hence requires a rhythmical structuring of its protracted everydayness. The solidarity and care expressed in a sustained participation in protective measures, on the one hand, was accompanied by the prominence of polyphonic calls to enjoy and productively fill leisure time or to use lockdowns to get fit or mindful, on the other (see also Anderson 2021a, 214). In view of those privileged enough to partly enjoy the pandemic as a retreat to the self, it cannot but strike the eye how the value of living-in-the-present simultanously defined governmental pandemic response and emerged as a yardstick of neoliberal technologies of the self. Perhaps we are already wittnessing an incipient internalisation of rhythmical living.

## Conclusion

Departing from the observation of the distinct place that fatigue occupies in (governmental) descriptions of pandemic life, this chapter argued that analysing its problematisation allows us to engage with the dynamic adaptation of the emergency imaginary to pandemic circumstances. The pandemic present suggests that a heuristic of states of exception has overestimated the capacity of state measures to govern the pandemic and undervalued the challenge that the temporal emergence of the pandemic, illustrated by non-linearity and exponentiality, poses. Emergencies, on the other hand, have long been negotiated as problems of governing the future. In this context, the pandemic causes a refocusing of governmental apparatuses on the (emergent) present and puts the question of the blurring of the emergency and the everyday at the centre of governance. Under pandemic conditions, the question of the emergency poses itself less in the register of a hope for a quick return to normality and instead points to a need of enduring a continuously extended present. Undermining the plausibility of imaginaries of governance that build their appeal on a progressively unfolding horizon towards which subjects aspire, the experience of this temporal suspension puts the availability of hope itself into question.

The emergence of 'pandemic fatigue' as a term employed to capture an experience of living through a protracted emergency thus signals a governmental concern for hope's loss of affect and that needs to be strategised in order to manage a present that is experienced as horizonless, recalling Claire Colebrook's observation that '[a]pproaches to apocalyptic transformation have become managerial' (Colebrook 2020, 179). In this sense, the entrance of fatigue into imaginaries of governance during the COVID-19 pandemic anticipates the Anthropocenic challenge of living exposed to nonhuman emergences and attunes them to a management of the affective implications of the intertwined lack of hope and horizon. In confronting this challenge, communicating through hope in order to restore the feelings of self-efficacy – needed to endure a present devoid of the clear temporal markers of a horizon and cultivating time-producing, rhythmised forms of subjectivity – becomes a prime governmental task.

But can there be hope in fatigue? From the analytical stance taken in this chapter, pandemic fatigue is an object of governmental thought that is being engaged with through managerial apparatuses. More pragmatically hopeful readings may suspect that fatigue's detachment could be a sign of a declining hubris of control, an affirmation of sorts of nonhuman temporalities, agents and emergences that the COVID-19 pandemic pointed to, and that the climate crisis undoubtedly re-rehearses. However, if being hopeful is increasingly appropriated by governments in the service of mere endurance, and what is stylised as an outside to power reliably turns out to be a fold, the only hope may subsist in its negation.

## Notes

1. Whilst not the main focus of this chapter, the presentist focus potentially depoliticises the fact that the structural conditions (within which the present unfolds) are the result of neoliberal reforms in the (public) health sector, reforms based on the continuous questioning of the need for excess capacities in the name of efficiency and profitability.
2. I take this idea from Luis Lobo-Guerrero who shared it with me during conversation.

## References

Adey, P., Anderson, B. and Graham, S. 2015. Introduction: Governing emergencies: Beyond exceptionality. *Theory, Culture & Society* 32(2): 3–17.

Agamben, G. 1998. *Homo Sacer: Sovereign Power and Bare Life.* Stanford, CA: Stanford University Press.

Agamben, G. 2005. *State of Exception.* Chicago, IL: University of Chicago Press.

Agamben, G. 2021. *Where Are We Now? The Epidemic as Politics.* London: Rowman & Littlefield.

Anderson, B. 2006. Becoming and being hopeful: Towards a theory of affect. *Environment and Planning D: Society and Space* 24(5): 733–52.

Anderson, B. 2014. *Encountering Affect: Capacities, Apparatuses, Conditions.* Farnham: Ashgate.

Anderson, B. 2016. Governing emergencies: The politics of delay and the logic of response. *Transactions of the Institute of British Geographers* 41(1): 14–26.

Anderson, B. 2017. Emergency futures: Exception, urgency, interval, hope. *The Sociological Review* 65(3): 463–77.

Anderson, B. 2021a. Affect and critique: A politics of boredom*. *Environment and Planning D: Society and Space* 39(2): 197–217.

Anderson, B. 2021b. Scenes of emergency: Dis/re-assembling the promise of the UK emergency state. *Environment and Planning C: Politics and Space* 39(7): 1356–74.

Anderson, B., Grove, K., Rickards, L. and Kearnes, M. 2020. Slow emergencies: Temporality and the racialized biopolitics of emergency governance. *Progress in Human Geography* 44(4): 621–39.

Australian Psychological Society. 2020. *Managing Lockdown Fatigue*. Available at https://psychology.org.au/getmedia/74e7a437-997c-4eea-a49c-30726ce94cf0/20aps-is-covid-19-public-lockdown-fatigue.pdf (accessed 2 January 2024).

Azzolino, D. and Cesari, M. 2022. Fatigue in the COVID-19 pandemic. *The Lancet Healthy Longevity* 3(3): e128–9.

Baraister, L. 2017. *Enduring Time*. London: Bloomsbury Academic.

Bargués-Pedreny, P. 2019. Resilience is 'always more' than our practices: Limits, critiques, and skepticism about international intervention. *Contemporary Security Policy* 41(2): 263–86.

Berlant, L. 2011. *Cruel Optimism*. Durham, NC: Duke University Press.

Bratton, B. 2021. *The Revenge of the Real: Politics for a Post-Pandemic World*. La Vergne: Verso.

Chandler, D. 2018. *Ontopolitics in the Anthropocene: An Introduction to Mapping, Sensing and Hacking*. Abingdon and New York: Routledge.

Colebrook, C. 2020. What would you do (and who would you kill) in order to save the world? Dialectical resilience. In D. Chandler, K. Grove and S. Wakefield (eds), *Resilience in the Anthropocene: Governance and Politics at the End of the World*. London and New York: Routledge, 179–99.

Cooper, M. 2006. Pre-empting emergence: The biological turn in the war on terror. *Theory, Culture & Society* 23(4): 113–35.

Daston, L. 2021. Ground-zero empiricism. *Critical Inquiry* 47(S2): S55–7.

Deleuze, G. 1995. The exhausted. *SubStance* 24(3): 3–28.

Dillon, M. 2007. Governing terror: The state of emergency of biopolitical emergence. *International Political Sociology* 1(1): 7–28.

Dillon, M. and Lobo-Guerrero, L. 2009. The biopolitical imaginary of species-being. *Theory, Culture & Society* 26(1): 1–23.

Fassin, D. and Fourcade, M. 2021. Introduction: Exposing and being exposed. In D. Fassin and M. Fourcade (eds), *Pandemic Exposures: Economy and Society in the Time of Coronavirus*. Chicago, IL: HAU Books, 1–18.

Flexer, M. J. 2020. Having a moment: The revolutionary semiotic of COVID-19 [Version 1; Peer Review: 1 Approved, 1 Approved with Reservations]. *Wellcome Open Research* 5: 134.

Foucault, M. 2003. [1997] *Society Must Be Defended: Lectures at the Collège de France, 1975–76*. New York: Picador.

Gäckle, N. (2023). Governing pandemic fatigue: An international relations case of experiential biopolitics. *European Journal of International Relations* 29(4): 1–26.

Gemignani, M. and Hernández-Albújar, Y. 2022. Neoliberal and pandemic subjectivation processes: Clapping and singing as affective (re)actions during the Covid-19 home confinement. *Emotion, Space and Society* 43: 1–7.

Gorfinkel, E. 2012. Weariness, waiting: Endurance and art cinema's tired bodies. *Discourse* 34(2–3): 311–47.

Grove, K., Rickards, L., Anderson, B. and Kearnes, M. 2022. The uneven distribution of futurity: Slow emergencies and the event of COVID-19. *Geographical Research* 60(1): 6–17.

Grusin, R. 2010. *Premediation: Affect and Mediality After 9/11*. London: Palgrave Macmillan.

Habersaat, K. B. et al. 2020. Ten considerations for effectively managing the COVID-19 transition. *Nature Human Behaviour* 4(7): 677–87.

Han, B.-C. 2021. The tiredness virus. Covid-19 has driven us into a collective fatigue. *The* Nation. Available at https://www.thenation.com/article/society/pandemic-burnout-society/ (accessed 2 January 2024).

Hartog, F. 2022. *Chronos: The West Confronts Time*. New York: Columbia University Press.

Honig, B. 2009. *Emergency Politics: Paradox, Law, Democracy*. Princeton, NJ: Princeton University Press.

Hutchings, K. 2008. *Time and World Politics: Thinking the Present*. Manchester: Manchester University Press.

Kattago, S. 2021. Ghostly pasts and postponed futures: The disorder of time during the Corona pandemic. *Memory Studies* 14(6): 1401–13.

Koselleck, R. 2004. *Futures Past: On the Semantics of Historical Time*. New York: Columbia University Press.

Krasmann, S. and Hentschel, C. 2019. 'Situational awareness': Rethinking security in times of urban terrorism. *Security Dialogue* 50(2): 181–97.

Lakoff, A. 2007. Preparing for the next emergency. *Public Culture* 19(2): 247–71.

Latour, B. 2021. Is this a dress rehearsal? *Critical Inquiry* 47(S2): S25–7.

Lee, P. M. 2021. Introduction: Aspiration burnout. *October* (176): 3–6.

Lefebvre, H. 2013. [1992] *Rhythmanalysis: Space, Time and Everyday Life*. London and New York: Bloomsbury Academic.

Lisle, D. 2023. The stubborn habits of migration: Self-care as endurance. In N. Salazar and J. Scheerder (eds), *Contemporary Meanings of Endurance: An Interdisciplinary Approach*. Abingdon: Routledge, 21–39.

Lobo-Guerrero, L. 2016. *Insuring Life: Value, Security and Risk*. London and New York: Routledge.

Nabi, R. L. and Myrick, J. G. 2019. Uplifting fear appeals: Considering the role of hope in fear-based persuasive messages. *Health Communication* 34(4): 463–74.

NHS England. 2021. Fatigue. *Your COVID Recovery*. Available at https://www.yourcovidrecovery.nhs.uk/managing-the-effects/effects-on-your-body/fatigue/ (accessed 2 January 2024).

Nixon, R. 2011. *Slow Violence and the Environmentalism of the Poor*. Cambridge, MA: Harvard University Press.

Povinelli, E. 2011. *Economies of Abandonment: Social Belonging and Endurance in Late Liberalism*. Durham, NC: Duke University Press.

Povinelli, E. 2021. *Between Gaia and Ground: Four Axioms of Existence and the Ancestral Catastrophe of Late Liberalism*. Durham, NC: Duke University Press.

Prozorov, S. 2021a. A farewell to Homo Sacer? Sovereign power and bare life in Agamben's coronavirus commentary. *Law and Critique* (online first): no pagination.

Prozorov, S. 2021b. Foucault and the birth of psychopolitics: Towards a genealogy of crisis governance. *Security Dialogue* 52(5): 436–51.

Rabi, M., Samimian-Darash, L. and Hilberg, E. 2022. Encapsulation: Governing actual uncertainty in the coronavirus pandemic. *Sociology of Health & Illness* 44(3): 586–603.

Reid, J. 2023. On endurance. The politics of being between pain and boredom. In N. Salazar and J. Scheerder (eds), *Contemporary Meanings of Endurance: An Interdisciplinary Approach*. Abingdon: Routledge, 9–20.

Roberts, S. L. and Elbe, S. 2017. Catching the flu: Syndromic surveillance, algorithmic governmentality and global health security. *Security Dialogue* 48(1): 46–62.

Ruwet, C. 2021. Par-delà les temps qui courent: Comment la pandémie de Covid-19 nous invite à refonder notre rapport au temps. *Revue de la régulation* 29: no pagination.

Salazar, N. and Scheerder, J. (eds). 2022. *Contemporary Meanings of Endurance: An Interdisciplinary Approach*. London: Routledge.

Samimian-Darash, L. 2022. Governing the future through scenaristic and simulative modalities of imagination. *Anthropological Theory* 22(4): 393–416.

Schlegel, L. 2022. Between climates of fear and blind optimism: The affective role of emotions for climate (in)action. *Geographica Helvetica* 77(4): 421–31.

Smith, J. E. 1969. Time, times, and the 'right time'; chronos and kairos. *Monist* 53(1): 1–13.
Stäheli, U. 2014. Hoffnung als ökonomischer Affekt. In I. Klein and S. Windmüller (eds), *Kultur der Ökonomie*. Bielefeld: transcript Verlag, 283–300.
Voelkner, N. 2019. Riding the shi: From infection barriers to the microbial city. *International Political Sociology* 13(4): 375–91.
Weber, S. 2022. *Preexisting Conditions Recounting the Plague*. New York: Zone Books.
WHO. 2020a. *COVID-19 Global Risk Communication and Community Engagement Strategy, December 2020 – May 2021: Interim Guidance, 23 December 2020*. Geneva: World Health Organization.
WHO. 2020b. *Mental Health Considerations during COVID-19 Outbreak*. Geneva: World Health Organization.
WHO. 2020c. *Pandemic Fatigue. Reinvigorating the Public to Prevent COVID-19. Policy Framework for Supporting Pandemic Prevention and Management. Revised Version November 2020*. Copenhagen: World Health Organization. Regional Office for Europe.
WHO. 2020d. Statement – Rising COVID-19 Fatigue and a Pan-Regional Response. Available at https://www.who.int/europe/news/item/06-10-2020-statement-rising-covid-19-fatigue-and-a-pan-regional-response (accessed 2 January 2024).
WHO. 2020e. *Survey Tool and Guidance. Rapid, Simple, Flexible Insights on COVID-19. Monitoring Knowledge, Risk Perceptions, Preventive Behaviours and Trust to Inform Pandemic Outbreak Response*. Copenhagen: World Health Organization. Regional Office for Europe.
WHO. 2021. 28 January 2021 – Statement from the Regional Director Henri Kluge. Available at https://www.youtube.com/watch?v=lAroEuuCQF0 (accessed 2 January 2024).
Wrangel, C. T. 2019. Biopolitics of hope and security: Governing the future through US counterterrorism communications. *Globalizations* 16(5): 664–77.
Zebrowski, C. 2019. Emergent emergency response: Speed, event suppression and the chronopolitics of resilience. *Security Dialogue* 50(2): 148–64.

# 7

# Securing the Hopeful Subject? The Militarisation of Complexity Science and the Limits of Decolonial Critique

*Claes Tängh Wrangel*

## Introduction

Across posthumanist critiques, hope is often claimed to be located beyond the human, *in* the world. In contrast to modernist conceptualisations of hope as a human prerogative, or as a particular psychological disposition,[1] the posthuman tradition commonly equates hope with the notion of emergence, as made possible by a complex world of radical contingency and relational entanglement. Such conceptualisations place modernity at war with hope. According to decolonial thinker Sylvia Wynter, reliance on objective science and linear causality has rendered modern Man – or *Rational Man* in Wynter's terminology – incapable of recognising the hope and 'magic' of human and nonhuman life. To save hope, which she holds to be 'dried up' (1989, 638) by modernity, Wynter argues for the invention of a new science, a decolonial *scienta*, that would recognise, rather than deny, the reality of complexity. 'Hope, if it is to exist', Wynter writes, 'would have to be founded in a new order of knowledge' (1984, 51). With explicit reference to the complexity sciences, in particular neurobiology, Wynter argues that the establishment of this new order of knowledge would be the first step 'towards the human, after Man', enabling 'a new and more inclusive mode of "human nature"', beyond our 'present conflictive modes of group integration' (Wynter 2003, 257).

This chapter aims to problematise and contextualise the notion of posthuman hope *as* complexity, that Wynter perceives as ungovernable and decolonial by default (1984; see also McKittrick 2015, 156). It does this by examining how the US military has appropriated insights from the complexity sciences – including the complexity sciences' critique of modernity's reliance on bounded reason and unsituated knowledge – as technologies of security and war well beyond the traditional battlefield. In sharp contrast to Wynter, the US military perceives complexity science as '*the* science of power' (NSI 2015, 46, emphasis added), a key to govern and secure 'vulnerable' (NSI 2013: 67) populations in an increasingly complex and volatile world.

To analyse how insights from the complexity sciences are politicised and militarised, this chapter examines the US Department of Defence's Strategic Multilayer Assessment (SMA) programme, a US inter-agency research network sponsored and facilitated by the US Office of the Secretary of Defense and the Joint Staff, with the official objective to manage 'risks associated with rapid change under conditions of fundamental uncertainty' (NSI 2023).[2] Within the SMA programme, the complexity sciences have been taken to represent a hope for a more sustainable form of security governance, an opportunity to influence the complex processes of becoming that is taken to define life, as well as the world we inhabit.

Central to this ambition is an explicit reformulation of the human subject – from Rational Man to a figure of life that could perhaps best be referred to as Complex Man: relational, potential, unpredictable, yet whose emergence remains within, rather than challenges, the colonial, racial and economic exclusions characteristic of our political present. As argued by Lt Col. Dave Lyle, a frequent contributor to the SMA programme, complexity theory offers a 'key' (NSI 2012a, 202) to govern what in the posthuman tradition has been called hope: to understand and unlock 'the driving forces behind creation itself' (NSI 2012a, 202). Such words give credence to David Chandler's observation that discourses of global governance increasingly perceive complexity and contingency as 'resource[s] that enable the extension of governance into new realms of "real" complex life' (2014, 34).

In order to analyse the relationship between complexity and security governance, this chapter reads complexity (and its neurobiological tandem concept plasticity[3]) not as objective scientific concepts, with a fixed meaning attached to a pregiven political telos, but rather as discursive articulations, conditioned by particular social, cultural and political contexts and power relations (Chandler 2014, 22). Indeed, as Michel Alhadeff-Jones has shown, complexity has a history (2008), one that is inseparable from the histories of modern governmentality and war. According to Alhadeff-Jones, the neurosciences and the related discipline of cybernetics were central to the emergence of complexity theory. As several studies have shown, research on complexity was heavily funded by the US military (Dillon and Reid 2009; Lawson 2018), who saw in the complexity sciences, primarily neurobiology, a means to govern human mental states in order to perform war more effectively (Howell 2017; Moreno 2012). As such, the recognition of a complex world not only predates the emergence of the Anthropocene, it could also be seen as occurring *within* modernist epistemology and politics, not least war. In this respect, it is interesting to note that the etymological roots of the concept of cybernetics, which is so closely affiliated to both complexity and neurobiology, stems from the Greek *kubernetes*: the art of governing (Alhadeff-Jones 2008, 70).

The chapter will proceed as follows. In the next section I will engage how the relationship between complexity and governance has been articulated in decolonial critiques of modernity. I examine Wynter's work in depth, as she not only has been widely influential in the formation of decolonial posthumanist critique, but also explicitly draws links in her work between hope, complexity and decoloniality. The second, empirically oriented section, analyses the SMA programme's appropriation of complexity science and neurobiology in order to showcase how the politicisation and militarisation of complexity is imagined and practised today. I conclude the chapter by questioning what the appropriation of complexity by military discourses may come to mean for our hope of a decolonial epistemology and politics, beyond the current postcolonial condition.

## Decolonial Hope: Complexity Beyond Governance?

For Wynter and other posthuman thinkers, recognition of complexity has been translated into an 'ethicopolitical aspiration' (Thakkar 2020, 75), a belief that the concept of complexity holds the capacity to move beyond the racial, colonial and economic exclusions, not to mention the environmental degradation characteristic of modernity. In critical discussions of the Anthropocene, this hope is often located in the dissolution of the modern distinction between human and nature that the complexity sciences have been taken to represent. Elizabeth Grosz's description of the complex world as a site of 'wonder' (Grosz and Bell 2017), of 'creation or invention' (Grosz et al. 2017, 38) and of radical co-belonging (ibid.) is a case in point. According to Grosz, recognition of the reality of complexity is a radical political practice – a 'welcoming of the new' (Grosz et al. 2017, 130) – that would move life beyond the exclusions of modern biopolitics (137).

In that sense, complexity has been seen to firmly unsettle what Wynter has described as the figure of life that she deems to be over-represented or naturalised in modernity: Rational Man, defined by Wynter as bounded, individual and capable of objective knowledge (1984; 1989). According to Wynter, Rational Man was established during the Enlightenment, in a particular order of Western knowledge, which represented 'its own local culture and its conception of the human, as natural, supracultural and isomorphic with the human species' (1995, 17). The birth of the natural sciences is, according to Wynter, coterminous with the establishment of Rational Man: an 'organization of knowledge' (1984, 21; see also McKittrick 2015, 145) characterised by its attempt to objectively divide and categorise the world into distinct observable parts. Like many others (for example, Latour 2016), Wynter holds the use of binaries – between 'Sameness/Difference' (1984, 54), 'Nature/Culture' (1984, 27), 'Rational humans/Irrational animals' (1989, 316) 'Life/Death, Order/Chaos' (1984, 27) – to be exemplary of the modern desire for categorisation.

Through its reliance on objectivity, Wynter argues that modernity privileges the actual, that which can be observed, measured and objectified (1984, 21), in order to infer generalised patterns and law-like causal relations between distinct objects. Drawing on Wynter's work, Walter Mignolo has described this logic of science as a form of 'global linear thinking', an 'imperial scientia' that perceives 'connections through time, including epistemic breaks and paradigmatic changes, [as] follow[ing] one another in a linear fashion' (2015, 110; 117). In Wynter's words, modernity was institutionalised by an equation between linear thinking, causality and universal reason, establishing an 'authority of Reason, the Reason coded by the Natural Logos of humanism based on the explanatory principle of a Natural Causality verified by the truth of empirical reality' (1984, 33).

Importantly, Wynter sees the constitution of Rational Man as having been made possible through a biopolitical and ontopolitical war with the complexity and plasticity of life – privileging being and actuality over becoming and potentiality (Mignolo 2015, 119). The complex, contingent and social processes through which life comes into being, what Wynter refers to as the 'long processes of the self-making' (1984, 50), thus remain 'unseen' (Chandler in Chapter 1) by modernity, suppressed and abjected. According to Wynter, modern science 'repress[es] any awareness that 'causes come to us from many sides' and that humans live in a 'multi-linear, multi-reinforcing causal world' (Riedl with Kaspar 1984, quoted in Wynter 1984, 32). According to Wynter, it is this repression of complexity that makes hope all but impossible in modernity. Caught in actuality, modernity is held to be 'without any hope of being able to have any valid knowledge of reality except through the mediation of the very paradigms that excluded any such hope' (2003, 275–6). Elsewhere, she has described hope as being 'dried up' (1989, 638) by modernity, blinding Rational Man from seeing the world as 'an interacting system' (1984, 22) engaged in constant 'dynamic change', thus rendering impossible the emergence of 'a newer and still evolving world' (ibid.).

With reference to biologist Lewis Thomas, Wynter argues that 'hope, if it is to exist, would have to be found in a new order of knowledge', a 'post-atomic' (1984, 51) order that would be as 'solid' and 'hard as physics', yet which would recognise the 'ambiguities' that 'plague' (1984, 43) physical and biological reality. To change modernity, Wynter thus places her hopes in the formation of a new science, a decolonial *scienta* capable of moving beyond binary definitions of Man, in order to recognise the 'invention' (2003, 331) and 'alterability' (McKittrick 2015, 160) that Wynter holds to be definitional of both human and nonhuman life, including geological life. Despite multiple declarations that change must come from below, from liminal groups abjected by Rational Man (1984, 38; 49) – Wynter repeatedly equates and substantiates the notion of *scienta* by recourse to science established in the Global North, in particular neurobiology, as exemplified above. In her article 'The Ceremony Must be Found', Wynter argues that 'it is only with science, as Riedl and Kaspar (quoting Roman Sexl) observe, that there is ever any true 'victory over the ratiomorphic apparatus' (1984, 37).

The 'Human, after Man' (Wynter 2003) that this *scienta* is both to be based upon as well as to issue forth is described by Wynter as an embodiment of hope, as 'magical. *Bios* and *Logos*' (1995, 35). Employing terms from neurobiology – which as a discipline has sought to problematise the liberal division between mind and body, as well as between the self and the external world (Damasio 1994; Rose and Abi-Rached 2013; Whitehead et al. 2018) – Wynter argues that the life of hope – that is, 'real' life – is neither cultural or biological, but both simultanously: 'Words made flesh, muscle and bone animated by hope and desire' (1995, 35). Our ideas of who we are – 'our "worlds of mind" or modes of consciousness' (1984, 24) – are, according to Wynter, not simply an expression of our biological being, predetermined by our genes, but rather a dynamic mental state, the result of complex interactions between the neural brain and the social and physical environment. As such, neurobiology is taken not only to disrupt the modern binary between nature and culture - between 'language [and] pre-linguistic biological processes' (1984, 24) – but also to inscribe plasticity into the very definition of the human, portraying our brains not as pre-fixed entities but as open to external stimuli, mutable across our lives. As summarised by Katherine McKittrick, Wynter perceives life as a 'co-relational figure: [. . .] a figure who is *at once* physiologically organic, cognitively responsive, and creatively inventive and, *in this simultaneity*, provides the origin stories through which we make sense of our flesh-and-blood and neurological and cultural claims to humanness' (2015, 144, emphasis added).

Embodying both *bios* and *logos*, biology and culture, Wynter holds our images of Man to be essentially unfixed, open to change. The task of Wynter's *scienta*, is thus articulated as a calling to rewire, or 'redraw [. . .] in undared forms' (1995, 35) the neurobiological constitutions of our mind, to set loose 'unbounded or ungoverned brain activity' (McKittrick 2015, 156), thus 'consciously alter[ing] our mode of self-troping, [and the] rhetor-neuro-physiological program that constitutes our "world of mind"' (Wynter 1984, 52).

To set emergence – and hence hope – loose, is for Wynter a conscious activity, one that demands active work. With reference to Aimé Césaire, Wynter has called this new *scienta* a 'science of the word', drawing attention to Césaire's definition of language as a creative performance, capable of affecting the brain's right hemisphere, where Wynter finds the corporeal location of the imagination (1984, 51) and of our binary images of 'Self/Not Self' (ibid.). As described by McKittrick, the science of the word is a 'poetic' act, a 'creative labour', an 'interdisciplinary and collaborative task' that is taken to hold the potential to issue forth 'a new and more inclusive mode of "human nature"', beyond our 'present conflictive modes of group integration' (2015, 154).

While Wynter perceives the use of the brain as an act of resistance, as an activation of plasticity and emergence that escapes governance, others have argued convincingly that the concepts of plasticity and complexity do not in themselves exist outside of power, nor that they necessarily operate in opposition to modern forms of governance. Indeed, as several studies have shown, insights and techniques from the complexity sciences, including neurobiology, have been readily adopted by liberal democracies as a means of 'govern[ing] through [complexity] rather than "over" or "against" it' (Chandler 2014, 35; see also Dillon and Reid 2009; Rose and Abi-Rached 2013; Whitehead et al. 2018). To be sure, the governmental use and meaning attributed to complexity and neurobiology has varied greatly, stretching from practices aiming to *reduce* the plasticity of non-desired bodies (Ahuja 2020; Pitts Taylor 2019) to governmental orders geared towards *promoting* a kind of hyper-plastic body where the subject is formed into 'pure potentiality' (Iman Jackson 2020; see also Tängh Wrangel 2018; 2019), withheld from ever acquiring stable form. As a series of studies have shown, modes of governing *through* plasticity and complexity have often served to underwrite social inequality and reproduce global power relations, expressed for instance by neoliberal attempts to responsibilise vulnerable subjects (populations or individuals) to optimise their brain/body in the face of economic and social stress (Iman Jackson 2020; Mansfield 2012; Rose and Abi-Rached 2013; Whitehead et al. 2018).

Taken together, the variety with which plasticity and complexity is hailed into politics calls on us to remember, as argued by Thakkar, that 'plasticity is itself made plastic, put to work for political projects that draw recursively on its power. Plasticity has no inherent political meaning, but it can nonetheless be made to do political work' (2020, 88). In the next section, I will trace how the 'political work' of complexity and plasticity is articulated by the US military, paying particular attention to how the US Department of Defense – through the SMA programme – identifies in complexity new modes of governance, and new definitions of order and power.

**Militarising Complexity: Governing the Emergence of Complex Life**

Within the SMA programme, complexity theory is taken to represent a metatheory of change and order: a 'key to better understanding the driving forces behind creation itself' (NSI 2012a, 202) capable of unlocking 'the fundamental and universal nature of *dynamic change* in the real world' (ibid., 216). Breaking modernist binaries between human and nature, as well as conceptual and operational distinctions between technological and ecological systems, complexity theory is described as linking seemingly separate worlds under one explanatory framework. According to Lyle, 'whether we're talking about the assembly of chemical compounds, neurons in our brains forming ideas and storing memories, the evolution of biospheres, the generation of economies, and even to the formation of galaxies, the basic process of emergence is the same' (NSI 2012a, 223). So presented, complexity theory represents for the SMA programme a 'universal framework' (ibid., 202) that would allow the US military to study 'common traits' (ibid.) of systems traditionally deemed separate and apart, to 'crossover models, metaphors, and ideas from one scientific discipline to others' (ibid.). As Lyle says, complexity theory 'improv[es] the way we look at *everything*' (ibid., 215, original emphasis).

In explicit contrast to modernist attempts of categorising, breaking up and solving 'problems', associated within the SMA programme with 'Newtonian (i.e. linear) implications of connectedness', complexity is presented as an attempt to 'explain the whole as well as the parts' (NSI 2017b, 32), emphasising relations and 'non-equilibrium dynamics and structures' rather than linear causal relations between fixed properties. While a complex system is described as consisting of relatively simple building blocks – such as the neuron – the SMA programme emphasises that complexity theory focuses on the networked interaction between different parts. According to Lyle, taking the example of neural complexity, 'it is not the shape of the neuros, or the strength of the chemicals [that matter]. It is their interactive combination, their place within and relation to *neural networks*, that makes the emergent properties of memory and thought happen' (NSI 2013, 58, original emphasis).

In that sense, complexity is articulated as an alternative to 'our traditional [hierarchical and tree-like] view of structures and how we act on them (whether military, technological, social, political, etc.)' (NSI 2017b, 35). Instead of such 'reductionist method[s]' (NSI 2012a, 220), complexity science offers a networked view of the complex security environment as existing on multiple interconnected scales and levels: 'a hypergraph spreading across and through an amorphous medium' (NSI 2017b, 35). As hypergraphs, complex systems are held to be unpredictable, characterised by 'interacting dynamic states with a highly non-linear relationship' (NSI 2017b, 32) and capable of producing bottom-up emergent phenomena. As argued by Lyle, this has critical implications for how security governance is to be realised:

> A key implication of emergence is that you cannot just impose macro properties from the top down without the supporting microprocesses to sustain them – the driving force of creation are local interactions at the micro level responding to local rule sets, and abrupt changes to these processes are more likely to crash than control the system. (NSI 2013, 56)

In similar words to those of Wynter, who sees in complexity a world of wonder and creative emergence, the US military perceives complexity as a 'creative engine' (NSI 2012a, 228), capable of 'produc[ing] amazing novelty' (NSI 2013, 55) and 'innovation' (Wright 2021a, 27) as well as of generating self-organisation and order, with 'no central authority directing their actions' (NSI 2013, 57). However, complexity is also seen as capable of 'generating extreme events', unwelcome 'surprise' (NSI 2012a, 202), 'unpredictability' (NSI 2017b, 32) and 'promoting' (NSI 2017b, 210) 'fundamental uncertainty' (NSI 2012a, 211). Through complexity, insecurity is articulated as an ontological condition. 'Complete elimination of surprise' is thus seen within the SMA programme as 'an impossible goal' (NSI 2012a, 210). Indeed, even 'metastable' or 'deceptively stable' states are held to be potentially disruptive, 'fully capable of generating extreme events that will surprise decision makers' (NSI 2012a, 217).

Within this explanatory framework, threats are thus not defined (as per traditional liberal accounts) as external to civilisation, as emanating from those deemed hopeless (Milton et al. 2013), without reason (Zehfuss 2012), or from the unconnected (Reid 2009), but are rather seen as an auto-immune effect, internal to the system. Threats are perceived to emerge 'not out "of the blue", [but] from our own systems and their tendency to become more complex (i.e., systems-of-systems-of-...-of systems, where complexity grows exponentially, not linearly)' (NSI 2012a, 210). Drawing on the neurosciences, complexity and plasticity have likewise been seen as inscribing insecurity and contingency as the very nature of being: 'Humans [. . .] are not static' a 2021 SMA report argues, concluding that radicalisation is a perpetual potentiality (Wright 2021b, 9), a 'vulnerability' inherent in 'human societies' (ibid., 7). According to Lyle, 'violent extremists are human, just like the rest of us' (NSI 2013, 61).

In the future, increasingly interconnected complex systems – of social, human, technological and ecological kinds (NSI 2012a, 210) – are further seen to create compounded, accelerated and as yet unknown threats, so-called 'black swans' (NSI 2020). Boundaries, once perceived to be fixed, such as those between domestic and foreign, between friend and foe (Wright 2021a, 24), are now articulated as 'deeper' and 'lumpier' (ibid., 25), thereby establishing what is articulated as a '"seam" between the external and the internal' (Wright 2021b, 9). According to the conference proceedings of the 2015 SMA Annual Conference, 'threats and challenges will be trans-regional, multi-domain, and multi-functional' (NSI 2017a, 39), rendering the 'whole of society [. . .] a vast attack surface' (Wright 2021a, 26). In a 2017 SMA White Paper titled *Influence in an Age of Rising Connectedness* we are informed that 'our environment is now characterized by non-uniformity and starts, stops, and leaps across orders of magnitude, and across geographical areas and socio-economic-political sectors' (NSI 2017b, 37).

While this may read as an abandonment of governance, as an acknowledgement of the impossibility of control and as 'the polar opposite of what the military seeks to achieve' (NSI 2012a, 219), this is far from the case. Quite the contrary: in contrast to how complexity has been appropriated within critical and decolonial theory as an 'ethicopolitical aspiration' (Thakkar 2020, 75), subversive of racial, colonial and economic hierarchies and linear imaginaries, the US military sees in complexity a new vision of governance, not 'an obstacle, [but] the very thing that makes creating our desired outcomes possible' (NSI 2012a, 220).

Indeed, the SMA programme explicitly refers to complexity science as '*the* science of power' (NSI 2015, 46, emphasis added), a means to unlock 'power in its purest form' (NSI 2017b, 25). If what the SMA programme describes as the 'dynamic nature of the environment' (2017b, 32) is embraced, rather than feared and controlled, it is argued that complexity can be 'harnessed' (NSI 2012a, 227; NSI 2017a, 26; Wright 2021a, 12), employed to 'attack not just the symptoms of violent extremism, but the driving forces behind it as well' (NSI 2013, 54). According to Lyle, the complexity sciences give 'access to very big levers of influence, or help you avoid negative unintended ones, especially in systems so big you can never hope to control them through brute force alone' (NSI 2013, 56). Signifiers like 'shape' (NSI 2012a, 221), 'convert' (ibid.), 'guide' (2017b, 34), 'engineer' (ibid.), 'change the landscape' (ibid., 35), 'design' (2013, 54) and particularly 'influence' (2017a) describe what the complexity sciences are seen to offer the US military.

To govern the 'drivers' or 'attractors' (NSI 2012a, 223) of complex emergence, the SMA programme devotes particular attention to the neurosciences, which is referred to as 'a model for how to intervene in complex systems' (NSI 2013, 59). If emergence is seen to emanate from the bottom up, from 'local interactions at the micro level responding to local rule sets' (NSI 2013, 56), then the SMA programme defines these 'local rule-sets' as the 'networks of neurons in the heads of individuals' (NSI 2013, 64). According to Lyle, understanding and regulating the neural network of individuals is the first step in countering radicalisation: 'the foundations for all of what we are concerned about in military strategy ultimately resides in the physical world – ideas are networks of neurons in the heads of individuals' (NSI 2013, 64). Across the SMA programme's many publications, we find similar expressions of this line of reasoning, such as Jeannette Norden's stipulation that 'everything a human being does or thinks or feels is the result of neuron activity in the brain' (NSI 2011, 39) and James Old's assertion that 'the mind is an emergent property of the brain and [. . .] violence is an emergent property of the mind' (NSI 2010, 10). Through such words, definitions of complexity in the world are effectively transposed onto the very definition of life, firmly establishing a figure of Complex Man as central to practices and logics of security. Such formulations also redefine the liberal concept of will as based on metaphysical reason – Rational Man in Wynter's terminology – into a corporeal 'behaviour' (NSI 2017b, 28), a neurological firing pattern, open to regulation.

Like Wynter, neurobiology perceives the human mind as both *bios* and *logos*, constituted by a complex interplay between biology and culture. Our sense of self is perceived as a socially and neurologically mediated mental construction, amenable to external influence. According to William Casebeer, Senior Research Manager in Human Systems and Autonomy at Lockheed Martin and a member of the SMA programme from its inception, 'the brain is a social organ, providing many opportunities for influence' (NSI 2012b, 58). Perceived as both *bios* and *logos*, neurobiology is held within the SMA programme to *both* be 'influenced by, *and* influence individual and group beliefs, intentions, and patterns of psycho-social activities' (NSI 2012b, 48, emphasis added). Social norms and values – 'the plots of our legends, morality tales, religious codes, laws' (NSI 2013, 59) – are particularly emphasised as key sites of intervention. They represent 'the attractors of social behavior' (ibid.), affecting, for instance, 'memory, reasoning (emotion and cognition), and identity' (NSI 2010, 12).

Through its emphasis on socially mediated neural representations of the world as the site of bottom-up emergence, the SMA programme thus shifts attention from the external social, political and ecological environment as such – its exclusions, its inequalities and its biopolitical valuation of different forms of life – to how this world is represented in the human mind. Far from subverting global exclusion, complexity science is appropriated to render global inequalities manageable. The 2014 SMA conference proceedings are explicit on this account, arguing that successful interventions should effectively *not* target material factors such as poverty or social inequality, but target how these are handled by the human brain, thus rendering 'causal factors that contribute to violent mobilization [. . .] *less efficacious*' (2014, 54, emphasis added). While the correlation between social exclusion and 'contentious politics in general, and terrorism in particular' is recognised by the SMA programme, it is thus effectively abandoned as an appropriate site of intervention: '[material] causes have a proximate psychological mechanism – they exert influence by affecting the human mind/brain' (2018, 129–30). Rather than problems to be solved, material inequalities and social exclusions represent, for the SMA programme, opportunities for intervention. As stated by General Joseph Votel in his keynote at the 2015 SMA Conference, 'unaddressed grievances of populations are creating exploitable opportunities that enables Gray Zone activity' (NSI 2015, 28). Other members of the SMA community have argued that 'we need to decide where we intervene at the point of social inequality' (NSI 2015, 45). Elsewhere, the idea of eradicating root causes of violence has been articulated as a modernist utopia, born out of a 'deterministic belief in radicalization' (NSI 2012b, 60) as a linear process directed by single causes.

While the turn to complexity may appear to be a shift from human-centric forms of governance – disrupting the modern distinction between human and nature – the above reading shows that complexity science's appropriation by the US military rather appears to consolidate the human-centrism emblematic of modernism. Politics is redefined into a question of human perception, a question of managing the poor rather than challenging the means of their exclusion. The means listed by the SMA programme to this end include, for instance, 'real-time or near-real-time systems for monitoring health [. . .] in refugee camps worldwide' (NSI 2012a, 209), the 'manipulation' of medical markets in areas with 'the most vulnerable populations in terms of geography or physical capacity' (2013, 67), as well as the use of neurobiologically informed counternarratives designed to target specific areas of the brain among select populations (2013, 62). Such practices arguably transpose complexity and its emphasis on self-organisation, novelty, creation and unpredictability from its proclaimed original ontological status as 'universal law' (NSI 2012a, 202) to a bio- and ontopolitical project – a particular societal organisation that has to be carefully produced, monitored, managed and secured. If there is any hope in complexity, then its appropriation by the US military represents a form of hope that could only be described as stripped of political potency, one that is employed not to release, but to regulate political agency; not to welcome the new, but to govern the future – rendering the future a mirror of the present.

## Conclusion: Beyond Rational Man, towards Complex Man

The reading above problematises Wynter's view of the complexity sciences as a form of decolonial *scienta of hope*, a 'new order of knowledge' (Wynter 1984, 51), capable of transforming Enlightenment discourse as well as the exclusions constitutive of modernity. What appears, in contrast, is what Mignolo would call an imperial *scienta* – a 'science of power' (NSI 2015, 46) – through which the brains and bodies of those deemed different and dangerous become subject to the modern gaze, reduced to objects of study and governance. Rather than acting as a force for hope, moving life 'towards the human, after Man' (Wynter 2003, 257), the US military's appropriation of complexity and neurobiology is perhaps better described as a reconfiguration of the colonial matrix around a complex and neurobiological subject. The figure of Rational Man replaced with a notion of Complex Man, seen as relational, open, creative and hopeful, yet equally excluding, equally controlling, equally universalised. Through complexity science, it is as if modernity itself is made plastic. If we do not see this, I contend, we risk repeating that which we seek to critique, thus placing our hope in the very structures and concepts that, to paraphrase Wynter, may make hope impossible.

## Acknowledgement

This research was supported by funding from the Swedish Research Council (2019-00683).

## Notes

1. See Tängh Wrangel (2018, 7, 24–6) for a discussion on hope in relation to liberal and modernist ideology.
2. Since its inception in the mid-2010s, the SMA programme has supported, conducted, coordinated and disseminated multidisciplinary research on the relation between complexity and security. In 2017, it was estimated that the SMA programme comprised 'around 3000 individuals, 95 universities, 14 US defense groups, and 8 foreign military groups' (NSI 2017a, 4).

3. In the neurosciences, plasticity refers to the idea that the human brain is an open and adaptable organ, mutable throughout life, formed through complex interrelations with the external environment. Etymologically, the concept derives from the Greek *pfassein*, meaning to mould. According to Malabou, 'the word *plasticity* has two basic senses: it means at once the capacity to *receive form* (clay is called "plastic", for example) and the capacity to *give form* (as in the plastic arts or in plastic surgery)' (2008, 5).

## References

Ahuja, N. 2020. Reversible human: Rectal feeding, plasticity, and racial control in US carceral warfare. *Social Text* 38(2): 19–47.

Alhadeff-Jones, M. 2008. Three generations of complexity theories: Nuances and ambiguities. *Educational Philosophy and Theory* 40(1): 66–82.

Chandler, D. 2014. *Resilience: The Governance of Complexity*. London: Routledge.

Damasio A. 1994. *Descartes Error: Emotion, Reason and the Human Brain*. New York: Avon Books.

Dillon, M. and Reid, J. 2009. *The Liberal Way of War: Killing to Make Life Live*. Abingdon: Routledge.

Grosz, E. and Bell, V. 2017. 'The Incorporeal': An interview with Elizabeth Grosz. *Theory, Culture & Society*. 22 May. Available at https://www.theoryculturesociety.org/blog/interviews-elizabeth-grosz-the-incorporeal (accessed 3 January 2024).

Grosz, E., Yusoff, K. and Clark, N. 2017. An interview with Elizabeth Grosz: Geopower, inhumanism and the biopolitical. *Theory, Culture & Society* 34(2–3): 129–46.

Howell, A. 2017. Neuroscience and war: Human enhancement, soldier rehabilitation, and the ethical limits of dual-use frameworks. *Millennium: Journal of International Studies* 45(2): 133–50.

Iman Jackson, Z. 2020. *Becoming Human: Matter and Meaning in an Antiblack World*. New York: New York University Press.

Latour, B. 2016. Onus orbis terrarum: About a possible shift in the definition of sovereignty. *Millennium: Journal of International Studies* 44(3): 305–20.

Lawson, S. T. 2018. *Nonlinear Science and Warfare: Chaos, Complexity and the U.S. Military in the Information Age*. London: Routledge.

Malabou, C. 2008. *What Should We Do with Our Brain?* New York: Fordham University Press.

Mansfield, B. 2012. Race and the new epigenetic biopolitics of environmental health. *BioSocieties* 7: 352–72.

McKittrick, K. (ed.). 2015. *Sylvia Wynter: On Being Human as Praxis*. Durham, NC and London: Duke University Press.

Mignolo, W. D. 2015. Sylvia Wynter: What does it mean to be human?. In K. McKittrick (ed.), *Sylvia Wynter: On Being Human as Praxis*. Durham, NC and London: Duke University Press, 106–23.

Milton, D., Spencer, M. and Findley, M. 2013. Radicalism of the hopeless: Refugee flows and transnational terrorism. *International Interactions* 39(5): 621–45.

Moreno, J. 2012. *Mind Wars: Brain Science and the Military in the 21st Century*. New York: Bellevue Literary Press.

NSI. 2010. *The Neurobiology of Political Violence: New Tools, New Insights. A Strategic Multilayer Assessment Workshop 1–2 December*. Available at https://nsiteam.com/proceedings-from-the-neurobiology-of-political-violence/ (accessed 3 January 2024).

NSI. 2011. *Strategic Multilayer Assessment (SMA), 5th Annual Conference, 29–30 November*. Available at: https://nsiteam.com/social/wp-content/uploads/2016/01/SMA-Annual-Conference-2011.pdf (accessed 3 January 2024).

NSI. 2012a. *National Security Challenges: Insights from Social, Neurobiological, and Complexity Sciences*. Available at https://nsiteam.com/insights-from-social-neurobiological-and-complexity-sciences/ (accessed 3 January 2024).

NSI. 2012b. *A World in Transformation: Challenges and Opportunities: Strategic Multilayer Assessment, 6th Annual Conference, 6–8 November*. Available at https://nsiteam.com/social/wp-content/uploads/2016/01/A-World-in-Transformation-Challenges-and-Opportunities-6th-Annual-Conference.pdf (accessed 3 January 2024).

NSI. 2013. *Strategic Multi-layer Assessment Periodic White Paper: Topics for Operational Considerations: Insights from Neurobiology and Neuropsychology on Influence and Extremism – An Operational Perspective*. Available at https://nsiteam.com/social/wp-content/uploads/2016/01/Insights-from-Neurobiology-Neuropsychology-on-Influence-and-Extremism—An-Operational-Perspective-April-2013.pdf (accessed 3 January 2024).

NSI. 2014. *8th Annual Strategic Multi-Layer Assessment (SMA) Conference 28–29 October 2014: A New Information Paradigm? From Genes to Big Data and Instagram to Persistent Surveillance: Implications for National Security*. Available at https://nsiteam.com/8th-annual-sma-conference/ (accessed 3 January 2024).

NSI. 2015. *9th Annual Strategic Multi-Layer Assessment (SMA) Conference 28–29 October 2015: No War/No Peace: A New Paradigm in International Relationship and a New Normal*. Available at https://nsiteam.com/9th-annual-sma-conference/ (accessed 3 January 2024).

NSI. 2017a. *10th Annual Strategic Multi-Layer Assessment (SMA) Conference. Jointly held with DHS, NCTC, DNI/NIC. From Control to Influence? A View of – and Vision for – the Future. Joint Base Andrews, 25–26 April.* Available at https://nsiteam.com/social/wp-content/uploads/2017/06/U_Final_SMA-Conference-Proceedings-25-26-April-2017.pdf (accessed 3 January 2024).

NSI. 2017b. *White Paper on Influence in an Age of Rising Connectedness. A Strategic Multilayer Assessment (SMA) Periodic Publication.* Available at https://info.publicintelligence.net/SMA-InfluenceConnectedness.pdf (accessed 3 January 2024).

NSI. 2018. *SMA White Paper: What Do Others Think and How Do We Know What They Are Thinking?* Available at https://nsiteam.com/social/wp-content/uploads/2018/03/White-Paper_What-Do-Others-Think_March2018_FINAL.pdf (accessed 3 January 2024).

NSI. 2020. *Key Factors Affecting Black Swans and Gray Rhinos in the USCENTCOM AOR: Expert Elicitations and Background Research.* Available at https://nsiteam.com/social/wp-content/uploads/2020/03/NSI_Key-Factors-Affecting-Black-Swans-and-Gray-Rhinos-in-the-USCENTCOM-AOR_March2020_Final.pdf (accessed 3 January 2024).

NSI. 2023. Strategic Multilayer Assessment. Available at https://nsiteam.com/sma-description/ (accessed 3 January 2024).

Pitts-Taylor, V. 2019. Neurobiologically poor? Brain phenotypes, inequality, and biosocial determinism. *Science, Technology, & Human Values* 44(4): 660–85.

Reid, J. 2009. Politicizing connectivity: Beyond the biopolitics of information technology in international relations. *Cambridge Review of International Affairs* 22(4): 607–23.

Rose, N. and Abi-Rached, J. (2013) *Neuro: The New Brain Sciences and the Management of the Mind.* Princeton, NJ: Princeton University Press.

Tängh Wrangel, C. 2018. *The Use of Hope: Biopolitics of Security furing the Obama Presidency.* Gothenburg: University of Gothenburg.

Tängh Wrangel, C. 2019. Biopolitics of hope and security: Governing the future through US counterterrorism communications. *Globalizations* 16(5): 664–77.

Thakkar, S. 2020. The reeducation of race: From UNESCO's 1950 Statement on Race to the postcolonial critique of plasticity. *Social Text* 38 (2 (143)): 73–96.

Whitehead, M., Jones, R., Lilley, R., Pykett, J. and Howell, R. 2018. *Neuroliberalism: Behavioural Government in the Twenty-First Century.* Abingdon: Routledge.

Wright, N. D. 2021a. *The Future Character of Information in Strategy: Forged by Cognition and Technology. Report for the Pentagon Joint Staff Strategic Multilayer Assessment Group.* Washington, DC: U.S. Department of Defense Joint Staff.

Wright, N. D. 2021b. *Cognitive Defense of the Joint Force in a Digitizing World. Report for the Pentagon Joint Staff Strategic Multilayer Assessment Group.* Washington, DC: U.S. Department of Defense Joint Staff.

Wynter, S. 1984. The ceremony must be found: After humanism. *boundary 2* 12/13(1–2): 19–70.

Wynter, S. 1989. Beyond the word of man: Glissant and the new discourse of the Antilles. *World Literature Today* 63(4): 637–48.

Wynter S. 1995. The Pope must have been drunk, the king of Castille a madman: Culture as actuality, and the Caribbean rethinking modernity. In A. Ruprecht and C. Taiana (eds), *Reordering of Culture: Latin America, the Caribbean and Canada in the 'Hood.* Ottawa, ON: Carleton University Press, 17–41.

Wynter, S. 2003. Unsettling the coloniality of being/power/truth/freedom: Towards the human, after man, its overrepresentation – an argument. *The New Centennial Review* 3(3): 257–337.

Zehfuss, M. (2012). Contemporary Western war and the idea of humanity. *Environment and Planning D: Society and Space* 30(5): 861–76.

# 8

# The Hope–Colonialism Nexus

*Marjo Lindroth and Heidi Sinevaara-Niskanen*

**Introduction**

The time we live in is marked by a search for hope. While humanity and the planet as a whole face near insurmountable challenges, we are regularly enjoined to retain hope. Whether the task ahead is maintaining faith in human rights (Sikkink 2017), finding solutions to environmental problems (Raygorodetsky 2017) or securing the future of democracy (Colomer and Beale 2021), solutions seem to always depend on the major role played by hope. The common perception is that hope is a transformative and progressive force, one enabling a better future. What is more, one sees hope playing an increasingly important role in political and social relations that have historically been devoid of hope, such as the relations between Indigenous peoples and states (Dahl 2012; Lightfoot 2016).

It is this forward-leaning appearance which this chapter examines, to reveal the ways in which this aspect of hope might camouflage its more regressive side. By this we mean the capacity of hope to stagnate progress and, in effect, paradoxically, to thwart hoped-for improvements. This facet of hope and its power implications have rarely been acknowledged in popular perception or research. By discussing how power is embedded in hope and the expectation to be hopeful, this chapter critically investigates the potential of hope to facilitate less noble ends. As we see it, hope is often complicit in creating and maintaining unequal and exploitative conditions; hope is a powerful political tool, regardless of the purposes to which it is put.

Specifically, this chapter brings together the discussions on contemporary colonialism and the many facets of hope, pointing to a connection that to date has largely remained under the radar of research. We ask: to what extent does hope lend itself to sustaining and extending colonial politics? It is no exaggeration to describe the current relations between Indigenous peoples and states as hopeful. Increased recognition of Indigenous peoples' rights, political inclusion and states' acknowledgement of past wrongdoings have materialised in milestones such as land claim agreements, formal apologies and truth and reconciliation processes (see, for example, Cameron 2015; Lightfoot 2016; Niezen 2017). These legal and political developments have been taken as evidence of increased equality and justice in the relations between Indigenous peoples and states – and, as such, as signalling the end of colonial aspirations. Yet, critical postcolonial scholarship has highlighted the persistence of colonial structures despite the promise of improved relations (Alfred 2005; Betasamosake Simpson 2017; Coulthard 2014; Simpson 2014). As we argue, the prevalent spirit of hopefulness may, in fact, be part of an effort to solidify colonial relations rather than dismantle them.

**Hope in a World of Uneven Relations**

In the midst of the changing conditions buffeting the environment, security and the economy, among other concerns, hope has become a sought-after human resource. Hope and the role that it plays in society has, as this volume demonstrates, sparked a growing social scientific interest in its sources, potential and power (Lindroth and Sinevaara-Niskanen 2021). The multiple political and social functions of hope have been examined in studies spanning resource extraction (Guerrieri 2019; Sejersen 2020), human rights (Ottendörfer 2019), nationalism (Hage 2003), securitisation (Anderson 2016; Tängh Wrangel 2019) and Indigenous rights (Lindroth and Sinevaara-Niskanen 2022; Miyazaki 2004). Hope plays many roles depending on the context (see Lindroth and Sinevaara-Niskanen 2021) and scholarship dealing with hope agrees on its growing pertinence. Drawing on insights into the links between hope and migration, Nauja Kleist and Dorte Thorsen (2017) describe our era as one that is epitomised by hope. The analysis of hope as affect has inspired a line of research, including the sources and essence of hope (Anderson 2016; Hage 2003; Lear 2008) and how hope can be fostered and maintained (Appadurai 2013; Miyazaki 2004). Hope has been investigated as an individual and collective affect (Massumi, 2015) as well as a societal and institutional expectation (Hage 2003; Nuijten 2004). Studies focusing on kinship concepts such as desire, optimism and confidence have also inspired research on hope (Berlant 2011).

In popular parlance, and research as well, hope tends to be understood as an empowering capacity (Lear 2008; Miyazaki 2004). The common perception of such a capacity – the ability to hope and to retain one's hopefulness – is that it is the same for everyone and can be readily summoned, especially in the face of uncertainty and challenge. Critical discussions on hope and development have, however, drawn attention to how hope is associated with larger questions of global (in)equality (see Chandler in Chapter 1). Arjun Appadurai (2007; 2013), who has studied globalisation processes, has pointed out how the meaning of hope is not universal. For example, for people living in the developing world, hope plays an important part in the ways in which social conditions are experienced. According to Appadurai, hope can facilitate the imagining of new kinds of horizons to come and, as such, may work to 'rectify' contemporary misery. Similarly, Kleist and Thorsen (2017), in work on African migration, highlight the centrality of hope for those struggling to improve their conditions. Indeed, it is imperative to acknowledge that the need for and significance of hope are not equally shared. The search for hope has a heightened role for those on the margins of society, whereas for the privileged and wealthy hope and its future horizons bear a different meaning.

This selectiveness – the way in which hope is distributed and allocated often on unequal terms – has intrigued scholars interested in hope as part of the contemporary exercise of power. Ghassan Hage (2003; 2016), in particular, has brought to the fore how hopes are a source of struggle in society. As he states, the already marginalised populations especially are expected to constantly 'dig for new forms of hope' (Hage 2003, 21). In a similar way, Lia Haro (2010), who has studied Indigenous peoples' position vis-à-vis the machinery of global development, draws attention to the disparities entailed in hope-laden discourses. As she observes, development projects repeatedly promise less deprivation, less hunger and more equal opportunity yet represent efforts that the privileged have undertaken to offer 'hope to those considered hopeless' (Haro 2010, 190).

Echoing these reflections, Valerie Braithwaite (2004) notes how the hopes of some groups prevail over the hopes of others, in priority as well as fulfilment. In other words, hope is selective in that those most in need of hope are burdened with the greatest expectations to increase their hopefulness. At the same time, their hopes and their efforts to remain hopeful may be ignored in favour of other types of hopes that have more economic or political value in society. In Haro's words, the latter are 'hegemonic hope[s]', ones that do not disturb the wealth and political position of the elite (Haro 2010, 191). As Ernst Bloch (1986) has put it, to hope is to reach towards a 'not-yet', to orientate to the future. As a vision of something yet to come, hope entails a risk or possibility that its object will never materialise (see Berlant 2011; Eagleton 2015). This risk of disappointment is higher for some than for others. It is this profound power disparity embedded in hope and hoping that we wish to delve into in what follows.

We argue that the focal question is not who has hope and who does not, but how much hope and what kinds of hopes are offered or made available to whom. This rationing of sorts is a question that goes beyond asking how hope is distributed to probing hope as a resource that can be utilised to many and various ends. Recasting the inclination to view hope as empowering, Ghassan Hage has drawn attention to how hope links with exploitation. Referring to the power relations between two groups, A and B, he points out that it is too simplistic to merely state that 'group A receives more hope than group B' (Hage 2016, 466). Instead, Hage elaborates a power dynamic that is more complex than mere distribution. As he continues, there is a need to recognise how 'group A extracts from group B, for itself, the very resources that group B needs to hope' (Hage 2016, 466). This 'process of extraction and exploitation' of hope creates the very foundation of inequality in society (Hage 2016, 466).

In another crucial insight, critical studies of humanitarian and health politics have observed a paradox whereby vulnerability forms a precondition for the 'right' to hope. The groups of people that are at the very margins of society – humanitarian immigrants in Europe, Indigenous communities struggling with suicide crises and victims of civil war or ethnic conflicts, among others – can only hope for recognition and care on the basis of demonstrable injury (see Allen 2021; Ottendörfer 2019; Stevenson 2014; Ticktin 2011). As hope is not available equally for everyone to begin with, it becomes a source and object of struggle.

Indeed, exploitation is embedded in the efforts to instil hope, whether these take place on an individual, collective or state level. Despite the growing scholarship on hope, this capacity of hope has remained largely undiscussed (for notable exceptions, see Warren 2015 on the position of Blacks within American politics and Berlant 2011 on 'cruel optimism' in society). As we see it, especially in the context of those previously marginalised – such as Indigenous peoples – it is more critical than ever to assess the ways in which hope(s) become rationed and how power is exercised through such practice. There is a politics at work in how hopes are rationed, that is, how hopes are meted out selectively. The rationing of hope entails not only offering hope selectively but also varying the hopes offered, both in degree and kind.

**Indigenising Hope**

Indigenous peoples are a particular marginalised group whose prospects have been and continue to be tied to hope. Historically, it could be said that it was critical for Indigenous peoples to hold on to their hope. Indeed, hope for the better is what has enabled the peoples to endure the intrusive measures and violence entailed in colonial practices. For example, Jonathan Lear's (2008) historical analysis of Indigenous peoples and hope emphasises their ability to retain their hope. By drawing on the history of the Crow Nation and the vanishing of the buffalo, Lear points out how, despite the destruction of their way of life, the Nation continued to hope for more than mere survival. This commitment to hope without the knowledge of what could be hoped for in their new situation amounted to what Lear terms 'radical hope' (see also Waldow in Chapter 11). Similarly, Haro (2010), in her analysis of the Indigenous communities that are part of the Zapatista movement, emphasises the way in which the peoples draw on hope and the long history of sustaining hope. As she notes, these communities 'situate themselves as members of an ongoing hopeful struggle of "longue durée"' and by doing so engage in a 'longer temporality of hoping and struggling rather than a temporality of despair, victimization and loss' (Haro 2010, 186). Indeed, Indigenous peoples have transformed what might seem like an unbroken history of hopelessness into a trajectory of radically hopeful resistance.

Taking the discussion on Indigenous peoples and hope to ongoing struggles, Ana Cecilia Dinerstein (2015) elaborates on the ways in which hope enables the restructuring of Indigenous autonomy. While several states have taken steps forward in recognising Indigenous peoples' rights and self-determination, the peoples are in many ways bound to state-centred processes. Citing developments in Argentina, Bolivia, Brazil and Mexico, Dinerstein notes how Indigenous peoples' particular capacity to hope also allows them to construct their own autonomy independently of state structures. As her research demonstrates, autonomy is 'the art of organizing hope' and a way to reach beyond the status quo (Dinerstein 2015). The accounts of Lear, Haro and Dinerstein emphasise hope's radical and transformative nature. Hope is what Indigenous peoples have and can rely on, even in the midst of colonial interventions by the state.

The exceptional connection between Indigeneity and hope has not gone unnoticed. Haro (2010) goes so far as to describe Indigenous peoples as experts in hope because, in their struggles, they have created and sustained their own hopes. These hopes may often differ radically from the institutional hopes proffered by colonial states or global development mechanisms. In research examining the role of hope in capitalist societies, Hage has also observed the distinct capacity of Indigenous peoples to find and retain hope. As he sees it, the rationing of hope – offering it to some and leaving some without it altogether – is one of the governing mechanisms tapped by neoliberal states. This social distribution, or withholding, of hope has led to what

Hage calls 'paranoid nationalism' – a state where 'no-hopers' experience 'a sense of entrapment, of having nowhere to go' (Hage 2003, 20; 21). In other words, Hage highlights how experiencing a 'scarcity' of hope pits different social groups against one another in a struggle for hope. In this context, as Hage (2003, 20) emphasises, those already 'overmarginalised populations of indigenous communities, homeless people, poor immigrant workers, and the chronically unemployed' are better equipped to cope with the loss of hope. Although Indigenous peoples' particular capacity to nurture hope might not change their social conditions, they are in some respects better positioned than those to whom the scarcity of hope is a new experience, such as 'urban dwellers paradoxically stuck in insecure jobs, farmers working day and night without "getting anywhere", small-business people struggling to keep their businesses going' (Hage 2003, 20).

Hage (2003) insightfully points out how Indigenous peoples' hope and hopefulness, borne of an extreme historical necessity, can now be harnessed by the peoples as a unique resource, an asset of sorts that other groups do not have. This is paradoxical in many ways, as Indigenous peoples have often been hoping against all odds. Even when the peoples and their way of life have faced fundamental crises, as Lear's historical analysis points out, their hope has persisted. In Lear's (2008, 96) words, it is 'a form of hope that seems to survive the destruction of a way of life'. This is where the radicality of Indigenous peoples' hope lies and how hope has worked to their advantage.

Ironically, Indigenous peoples have recently become a source of global hope. The growing concern for the environment and its degradation, in particular, have drawn attention to Indigenous peoples' traditional knowledge, their relationship to nature, and the ways in which such wisdom and attunement to the environment can foster hope for humankind. For example, Gleb Raygorodetsky (2017) discusses Indigenous cultures and livelihoods in terms of what they can offer the world and humanity. In his book *The Archipelago of Hope: Wisdom and Resilience from the Edge of Climate Change*, he portrays Indigenous peoples and their way of being in the world as sources of hope. As Raygorodetsky (2017) argues, Indigenous knowledge and ways of life represent radical, alternative engagements with the world, a type of environmental coexistence that is sorely needed. Studying the perceptions of indigeneity in contemporary culture and scholarship, David Chandler and Julian Reid (2019) also highlight how Indigenous peoples' attunement to the environment, once used as an impetus to 'develop' them, is now in demand. 'Becoming Indigenous' is presented as a source of hope and a solution for humankind and the natural world alike. These accounts not only emphasise the unique connection between Indigenous peoples and hope but also the problematics entailed in assuming that Indigenous peoples' hopes will facilitate change. In other words, Indigenous peoples and their cultures are perceived as leading the rest of us towards a more hopeful future – to the extent that we learn to indigenise our Western hopes and our ways of hoping.

While the entanglement of indigeneity and hope has gained increasing attention, the discussion has tended to be one-sided. Indigenous peoples are assigned the position of the hope-bearers and the rest of the world considers itself entitled to tap that capacity. Amidst celebrating Indigenous peoples' ability to hope, the question of whether and what kind of power is entailed in this engagement has remained unasked. What kinds of power are at work when Indigenous peoples are constantly hailed as hopeful and, as such, positioned as hope-givers?

There is no denying that the political and social position of Indigenous peoples has changed in recent decades and that this development has further encouraged hope (in all parties involved). From a broader perspective, however, Indigenous peoples' position in political participation, decision-making and rights-claiming has remained static. For example, a look at the inclusion of Indigenous peoples and their agendas in international politics reveals a two-fold development: institutions congratulate themselves for including Indigenous peoples yet all the while maintain structures that ultimately exclude them (Lindroth and Sinevaara-Niskanen 2018). Similarly, legal scholars have observed the reluctance of Western legal systems to fully recognise Indigenous peoples' rights (Birrell 2016; Young 2020). The recent reconciliation processes with states have also been criticised for being superficial and disappointing in their aims and outcomes from the perspective of Indigenous peoples (Niezen 2017; Regan 2011). In light of these discussions, there is a clear discrepancy between the developments that are taking place and the expectation that Indigenous peoples will retain their hope. One might well ask, what does this abundance of hope mean, in the case of Indigenous peoples, when hope is all that is on offer?

## Building Indigenous–State Relations

Colonialism, both historical and present-day, has marked the relations between Indigenous peoples and states. As Sanjay Seth (2013) notes, in research on international relations and postcolonialism, postcolonialism 'signifies the entire historical period after the beginnings of colonialism' (Seth 2013, 1). In a similar vein, Byron J. Good, May-Jo DelVecchio Good, Sandra Teresa Hyde and Sarah Pinto (2008, 6) describe the postcolonial as 'an era and a historical legacy of violence and appropriation carried into the present as traumatic memory, inherited institutional structures and often unexamined assumptions'. In sum, postcolonial scholarship has been committed to unveiling how colonial practices and mentalities persevere in the present. Within this line of enquiry, critical Indigenous studies have focused on the ways in which colonialism continues to reproduce marginalisation, violence and injustice for Indigenous peoples. This research has examined, among other topics, the shape-shifting nature of colonialism in Indigenous–state relations, the tactics of exclusive inclusion and the newly emerged political vocabularies that, in effect, continue past mentalities (Alfred 2005; Coulthard 2014; Simpson 2014; Stevenson 2014). Such analyses have pointed out how important it is to critically assess what lies beneath today's developments and their benevolent appearance.

Significantly then, postcolonial scholarship has demonstrated the enduring force, yet mercurial shape, of colonialism. As Lisa Stevenson (2014, 4) has aptly phrased it, 'our contemporary worlds are often haunted by the colonial in ways we do not fully understand'. In order to capture the covert nature of colonialism, Audra Simpson (2016a) utilises the concept of a 'grammar'. The term highlights how colonialism is embedded in societal patterns, which, like the grammar of a language, are rarely questioned. In her words, colonialism is:

> enduring, it has its own structure and logic and refusal as well, operating like a grammar and posture that sits through time. It is a politics deeply cognizant of its own production, of the never-ending nature of inequality and the need to stay the course. (Simpson 2016a, 329)

This idea of colonialism operating as an unquestioned grammar has led us to analyse the imbrication of hope within such structures of subordination. To date, postcolonial research has lacked attention to the critical analysis of hope and its connections to contemporary colonialism. This is not to say that Indigenous studies have failed to notice the increasing emphasis on hope and the problematics of hope and hopelessness, especially in Indigenous–state relations. Summarising the air of hopefulness that has entered these relations, Leanne Betasamosake Simpson (2017) describes how:

> It can appear or feel as if the state is operating differently because it is offering a slightly different process to Indigenous peoples. Goodness knows, we'd all like to feel *hopeful*. We'd like to see a prime minister smudging or acknowledging he is on Indigenous territory and have that signal a significant dismantling of settler colonialism [. . .] when the practices of settler colonialism appear to shift, it can appear to present an opportunity to do things differently, to change our relation to the state. (Betasamosake Simpson 2017, 45–6, italics added)

Betasamosake Simpson's observation brings into sharp relief the illusory nature of many of the hopes that animate Indigenous–state relations. As her critical account highlights, Indigenous peoples are highly aware of how elusive the promises are that have sprung from their allegedly improved relations with states.

With insights into the legacy of colonialism, Audra Simpson (2014; 2016a; 2017) has widely problematised the possibility of reciprocity between Indigenous peoples and states. As she has put it, 'the hope for mutuality (underwritten by a hope for sincerity)' offered by states might not be what Indigenous peoples seek as they recognise the failure of such gestures in the past (Simpson 2017, 29). Lisa Stevenson's (2014) research, for its part, reveals the challenge of achieving reciprocity even in the case of healthcare offered by states. She has studied Canadian welfare mechanisms aimed at tackling the tuberculosis epidemic in the past and addressing the current suicide crisis in Inuit communities. As she puts it, the benevolent and well-meaning interest of the state to care for Indigenous communities amounts to a 'murderous' intervention whereby the state acts according to its own ambitions with regard to welfare while ignoring the hopes of Indigenous communities.

With reference to the ongoing violence against and killing of Indigenous women and girls, in particular, Audra Simpson (2016b) fundamentally questions the trending idea of reconciliation that is held out as one of the main sources of hope in contemporary Indigenous–state relations. She reflects on the paradox in which the settler state expects Indigenous peoples to move on and forgive, yet all the while refuses to make any concessions (justice or land). Simpson's analysis raises an important question: what meaning does reconciliation have when a state is hostile towards Indigenous peoples and their political and legal systems, yet assumes that they are willing to invest their hopes in and engage with the state?

This elusiveness of hope comes also into play when states engage with other minority groups. Reflecting on the possibilities of freedom for Black peoples in contemporary state structures, Rinaldo Walcott (2021) invokes the notion of 'long emancipation', whereby Black peoples now have juridical emancipation but factual emancipation still awaits. In his words, the lives of Black peoples seem to abide in a condition where freedom 'is both belated and always just ahead of us' (Walcott 2021, 5; for a similar observation, see Warren 2015). Walcott's account of Black freedom as 'yet to come', situated in the field of African American studies, resembles the critique of state-centric systems in postcolonial and Indigenous studies. This critique has brought to the fore the requirement of waiting that characterises marginalised groups' struggles for justice. For example, Elizabeth Povinelli (2011, 128), in her work on late liberalism and colonialism, has pointed out the ways in which Indigenous peoples must 'persist in potentiality' while they wait for their rights and recognition (maybe endlessly). As we see it, hope has a significant role in sustaining this anticipatory state for all those seeking freedom and emancipation.

This leads us back to the question of how the desire to see Indigenous–state relations as hopeful works to sustain colonialism. As the accounts above demonstrate, waiting and hoping for something define the position of those who have been systematically marginalised and colonised. In this light, hope, which is all about expectation and faith in something to come, further cements the idea of Indigenous peoples, among others, as still having to wait. It is here, in first creating and then sustaining this structured stagnation, that hope acts as the grammar of colonialism today.

**The Colonial Time of Hope**

For Indigenous peoples, colonialism is an ongoing reality. Dispossession – the *raison d'être* of colonialism – is starkly evident in the lives of many Indigenous communities worldwide (Betasamosake Simpson 2017; Povinelli 2011; Regan 2011; Stevenson 2014). The aim of critical postcolonial scholarship is to understand the ways in which the ongoing social, political and legal dispossession of Indigenous peoples is maintained. Of particular interest here is how many of the envisioned and state-supported processes of rights and recognition have come to a halt and been left hanging as mere hopes. For example, Australia has seen a decades-long debate – with no resolution in sight – on amending the Constitution to recognise the Aboriginal and Torres Strait Islander peoples as Indigenous peoples, and to repeal its racist provisions. A similarly protracted process can be seen in Finland, where the state acknowledges the need to ratify the legally binding ILO Convention 169, yet continuously fails to do so. For the Sámi people in Finland, the Convention would provide strong legal backing in the discussion on cultural and land rights, among others (Lindroth and Sinevaara-Niskanen 2022). Even where recognition and reconciliation processes have been completed, the outcomes have often fallen woefully short of the hopes that ignited the processes in the first place. For example, the Truth and Reconciliation process in Canada, conceived to renew Indigenous–state relations, has been criticised for its limited scope, a shortcoming that enabled the state to evade full responsibility for justice and redress (Niezen 2017).

It seems that for those already marginalised the likelihood of hope going unfulfilled is higher than for other groups in society. Indigenous scholars have made similar observations and pointed to states making surface reforms – through legal recognition or political inclusion – and using various stalling tactics in order to evade Indigenous peoples' claims for change (e.g. Alfred

2005; Corntassel 2007; Simpson 2014). The present emphasis on hopefulness offers states fertile ground for postponing progress and diverting attention from the core issues. In keeping processes alive by constantly signalling hope, the promised improvements can be put off indefinitely.

As we see it, instead of being the driver of radical change, hope has worked to sustain colonial settings in Indigenous–state relations. In effect, what is at stake with the political utilisation of hope is the manoeuvring of time and temporality. Terry Eagleton (2015, 44) aptly notes how 'hope is the fetishism of the future, one that reduces the past to so much prologue and the present to mere empty expectancy'. Hope is a response that states can continuously and easily 'afford' when they are facing Indigenous peoples and their claims. It is as if the states are telling the peoples, with each case, that 'you need not wait quite so long this time', creating a perception that the requirement is less harsh in the light of the peoples' long history of persistent hoping.

Studying the Canadian Arctic and the impacts of colonialism on the Inuit way of life, Lisa Stevenson (2014) draws attention to the question of time. Focusing on the encounters between Indigenous peoples and the state, she highlights how the management of time was crucial for the colonial power in extending its regime. According to Stevenson, this meant that the Inuit had to be trained and disciplined to live by the coloniser's time. In her words (2014, 133), the Inuit way of life, which had depended on hunting, cooking and sleeping when the conditions were right, needed to adapt to the idea of time as 'an extractable resource' that could be used wisely or poorly. Time became a commodity and Indigenous peoples were to use it 'appropriately'. To us, this bears a striking resemblance to the contemporary workings of hope in Indigenous–state relations. As hope is always about futurity and not-yet-ness, it is always about time. Isn't it still the state's time that takes precedence over that of Indigenous peoples?

Taking her reflections on time further, Stevenson ends by asking: 'Is it possible that [. . .] territory is a minor trophy of colonialism – when compared to the domination of time?' (Stevenson 2014, 133). In light of the extensive cultivation of hope in the relations between Indigenous peoples and states, our response to this question is affirmative. In the era of rights and recognition, the colonial control of time has become less explicit but, as we argue, equally, if not more, powerful. While states might no longer discipline Indigenous peoples and force them to spend their time 'appropriately' or 'wisely', they have arrogated to themselves the right to ultimately define the value of time in political and legal processes that concern Indigenous peoples. Even more importantly, states retain the authority to decide whose time 'counts' and which matters are 'worth one's while'.

In sum, at the core of Indigenous–state relations today lies a hope–colonialism nexus. By actively creating and encouraging hopes, states shift the focus of these relations into the future. In other words, by default the colonial grammar of hope foregrounds the future, causing the present to recede from view. This is the way in which hope – despite its empowering appearance – works to stagnate and dilute Indigenous peoples' present, urgent, concerns.

## Conclusion

The increased cultivation of hope in the world today should alert our critical attention to the varying functions of hope in society. While hopes are abundant, it is crucial to see that the distribution and meaning of hope is not equal. Hopes may be rationed in a way that enables some people to hope at the expense of others. Indigenous peoples are – and have been – a group that has been enveloped by hope. There is historical evidence of Indigenous peoples' radical capacity to hope and to utilise that hope to resist colonial interventions. In the current environmental situation, Indigenous peoples, with their ways of being in the world, are even expected to provide hope for the rest of humanity. An air of hopefulness has also entered Indigenous–state relations, signalling the prospect of progress and improvements. However, critical postcolonial lenses lay bare links between hope and colonialism. Colonialism is a continuing practice and it has adopted a variety of forms throughout its existence. As we have argued in this chapter, the grammar underpinning colonialism today relies on hope. Hope, in essence, is about time and managing the trajectory of the future. Indeed, the relations between Indigenous peoples and states attest to a long history of Indigenous peoples' time being disciplined and labelled as unworthy. The emphasis on hope falls within a continuum of these colonial practices, in which states' prerogative in ruling over time has allowed them to nullify the urgency of Indigenous peoples' claims in the present. It is this control of time where hope's colonial potency lies.

## References

Alfred, T. 2005. *Wasáse: Indigenous Pathways of Action and Freedom*. Peterborough, ON: Broadview Press.

Allen, L. 2021. *A History of False Hope: Investigative Commissions in Palestine*. Stanford, CA: Stanford University Press.

Anderson, B. 2016. *Encountering Affect: Capacities, Apparatuses, Conditions*. New York: Routledge.

Appadurai, A. 2007. Hope and democracy. *Public Culture* 19(1): 29–34.
Appadurai, A. 2013. *The Future as Cultural Fact: Essays on the Global Condition*. London: Verso.
Berlant, Lauren. 2011. *Cruel Optimism*. Durham, NC: Duke University Press.
Betasamosake Simpson, L. 2017. *As We Have Always Done: Indigenous Freedom through Radical Resistance*. Minneapolis: University of Minnesota Press.
Birrell, K. 2016. *Indigeneity: Before and Beyond the Law*. Abingdon: Routledge.
Bloch, E. 1986. *The Principle of Hope*, vol. 1. Cambridge, MA: MIT Press.
Braithwaite, V. 2004. Preface: Collective hope. *The Annals of the American Academy of Political and Social Science* 592(1): 6–15.
Cameron, E. 2015. *Far Off Metal River: Inuit Lands, Settler Stories, and the Making of the Contemporary Arctic*. Vancouver, BC: UBC Press.
Chandler, D. and Reid, J. 2019. *Becoming Indigenous: Governing Imaginaries in the Anthropocene*. London: Rowman & Littlefield.
Colomer, J. M. and Beale, A. L. 2021. *Democracy and Globalization: Anger, Fear, and Hope*. New York: Routledge.
Corntassel, J. 2007. Partnership in action? Indigenous political mobilization and co-optation during the first UN Indigenous decade. *Human Rights Quarterly* 29(1): 137–66.
Coulthard, G. S. 2014. *Red Skin, White Masks: Rejecting the Colonial Politics of Recognition*. Minneapolis: University of Minnesota Press.
Dahl, J. 2012. *Indigenous Space and Marginalized Peoples in the United Nations*. New York: Palgrave Macmillan.
Dinerstein, A. C. 2015. *The Politics of Autonomy in Latin America: The Art of Organising Hope*. New York: Palgrave Macmillan.
Eagleton, Terry. 2015. *Hope without Optimism*. Charlottesville: University of Virginia Press.
Good, B., DelVecchio Good, M.-J., Hyde, S. T. and Pinto, S. 2008. Postcolonial disorders: Reflections on subjectivity in the contemporary world. In M.-J. DelVecchio Good, S. T. Hyde, S. Pinto and B. Good (eds), *Postcolonial Disorders*. Berkeley, CA: University of California Press, 1–40.
Guerrieri, V. 2019. The spatiality of hope: Mapping Canada's northwest energy frontier. *Globalizations* 16(5): 678–94.
Hage, G. 2003. *Against Paranoid Nationalism: Searching for Hope in a Shrinking Society*. Annandale, NSW: Pluto Press Australia.
Hage, G. 2016. Questions concerning a future-politics. *History and Anthropology* 27(4): 465–7.
Haro, L. 2010. The affective politics of insurgent hope. In J. Horrigan and E. Wiltse (eds), *Hope Against Hope: Philosophies, Cultures and Politics of Possibility and Doubt*. Amsterdam: Rodopi, 183–206.

Kleist, N. and Thorsen, D. (eds). 2017. *Hope and Uncertainty in Contemporary African Migration*. New York: Routledge.

Lear, J. 2008. *Radical Hope: Ethics in the Face of Cultural Devastation*. Cambridge, MA: Harvard University Press.

Lightfoot, S. 2016. *Global Indigenous Politics: A Subtle Revolution*. London: Routledge.

Lindroth, M. and Sinevaara-Niskanen, H. 2018. *Global Politics and Its Violent Care for Indigeneity: Sequels to Colonialism*. New York: Palgrave Macmillan.

Lindroth, M. and Sinevaara-Niskanen, H. 2021. Politics of hope: Transformation or stagnation? In A. H. Hosseini, B. K. Gills, J. Goodman and S. Motta (eds), *The Routledge Handbook of Transformative Global Studies*. Abingdon: Routledge, 230–42.

Lindroth, M. and Sinevaara-Niskanen, H. 2022. *The Colonial Politics of Hope: Critical Junctures of Indigenous–State Relations*. London: Routledge.

Massumi, B. 2015. *Politics of Affect*. Cambridge: Polity Press.

Miyazaki, H. 2004. *The Method of Hope: Anthropology, Philosophy, and Fijian Knowledge*. Stanford, CA: Stanford University Press.

Niezen, R. 2017. *Truth and Indignation: Canada's Truth and Reconciliation Commission on Indian Residential Schools*. Toronto, ON: University of Toronto Press.

Nuijten, M. 2004. Between fear and fantasy: Governmentality and the working of power in Mexico. *Critique of Anthropology* 24(2): 209–30.

Ottendörfer, E. 2019. Assessing the role of hope in processes of transitional justice: mobilising and disciplining victims in Sierra Leone's Truth Commission and Reparations Programme. *Globalizations* 16(5): 649–63.

Povinelli, E. 2011. *Economies of Abandonment: Social Belonging and Endurance in Late Liberalism*. Durham, NC: Duke University Press.

Raygorodetsky, G. 2017. *The Archipelago of Hope: Wisdom and Resilience from the Edge of Climate Change*. New York: Pegasus.

Regan, P. 2011. *Unsettling the Settler Within: Indian Residential Schools, Truth Telling, and Reconciliation in Canada*. Vancouver, BC: UBC Press.

Sejersen, F. 2020. Brokers of hope: extractive industries and the dynamics of future-making in post-colonial Greenland. *Polar Record* 56(E22).

Seth, S. 2013. Introduction. In S. Seth (ed.), *Postcolonial Theory and International Relations: A Critical Introduction*. London: Routledge, 1–12.

Sikkink, K. 2017. *Evidence for Hope: Making Human Rights Work in the 21st Century*. Princeton, NJ: Princeton University Press.

Simpson, A. 2014. *Mohawk Interruptus: Political Life Across the Borders of Settler States*. Durham, NC: Duke University Press.

Simpson, A. 2016a. Consent's revenge. *Cultural Anthropology* 31(3): 326–33.

Simpson, A. 2016b. The state is a man: Theresa Spence, Loretta Saunders and the gender of settler sovereignty. *Theory & Event* 19(4): 1–16.

Simpson, A. 2017. The ruse of consent and the anatomy of 'refusal': Cases from Indigenous North America and Australia. *Postcolonial Studies* 20(1): 18–33.

Stevenson, L. 2014. *Life Beside Itself: Imagining Care in the Canadian Arctic.* Oakland, CA: University of California Press.

Tängh Wrangel, C. 2019. Biopolitics of hope and security: Governing the future through US counterterrorism communications. *Globalizations* 16(5): 664–77.

Ticktin, M. 2011. *Casualties of Care: Immigration and the Politics of Humanitarianism in France.* Oakland, CA: University of California Press.

Walcott, R. 2021. *The Long Emancipation: Moving toward Black Freedom.* Durham, NC: Duke University Press.

Warren, C. L. 2015. Black nihilism and the politics of hope. *CR: The New Centennial Review* 15(1): 215–48.

Young, S. 2020. *Indigenous Peoples, Consent and Rights: Troubling Subjects.* New York: Routledge.

# 9

# Hopeful Times, Black Futures, and Things Quantum Technologies Tell about International Institutions

*Geoff Gordon*

**Cruel, Critical Optimism**

This edited collection starts from a premise that '[a]s traditional forms of governance and traditional hierarchies of power lose their relevance [in the Anthropocene], the politics of hope has increasingly come centre stage' (Waldow in Chapter 11). This chapter will cover these three things – traditional hierarchies of power, changing modes of governance, and politics of hope – by looking at international institutions through the lenses of divergent notions and programmes of hope, contrasting two possibilities for hope in the Anthropocene: on the one hand, cruel optimism; on the other, something like hope draped in black, or a fugitive hope that is 'mired in life' (Berlant 2011; Bey 2018; Winters 2016). I mean to identify the former with modern discourses of international law, both mainstream and critical. For the latter, I turn primarily to lessons from Black social theory. I will examine these different possibilities for hope by reference to two ambitious projects organised around imaginaries of quantum technology, namely Black Quantum Futurism, on the one hand, and the European Quantum Flagship on the other. Quantum technology and theory are contributing to and accelerating challenges to constraints of linear rationality in the Anthropocene, including for institutions of international relations and law, and invite more than one sort of radical hope.

Lately there have been moves towards a hopeful, critical foregrounding of contingency in international law (Venzke and Heller 2021). This chapter is informed by the understanding that reconstructions of how international law could be (or could have been) otherwise have served to stabilise socioeconomic relations characterised by unequal distributions of wealth and voice (Gordon 2021b). The reconstructive programme is premised upon consistent reproduction of authority internationally: demonstrating how international law could have been otherwise ultimately reinforces a belief that international law and international lawyers have the resources to get it right next time. In that sense, the situation loosely approximates elements of what Lauren Berlant describes in *Cruel Optimism*.

By optimism, Berlant means 'a sustaining inclination to return to the scene of fantasy that enables you to expect that this time, nearness to this thing will help you or a world to become different in just the right way' (2011, 2). By cruel, she means that 'the very pleasures of being inside a[n optimistic] relation have become sustaining regardless of the content of the relation, such that a person or a world finds itself bound to a situation of profound threat that is, at the same time, profoundly confirming' (2011, 2). Berlant focuses her inquiry on everyday affects to examine 'what thriving might entail amid a mounting sense of contingency' (2011, 11); it is both timely (van den Meerssche and Gordon 2020) and helpful to keep in mind that everyday affects and systemic pretensions are entangled and not wholly separable. From Berlant again: '*Cruel optimism* gives a name to a personal and collective kind of relation and sets its elaboration in a historical moment that is as transnational as the circulation of capital, state liberalism, and the heterofamilial, upwardly mobile good-life fantasy have become' (2011, 11).

Much of the foregoing approximates critiques of linear progress narratives in international relations and other fields (Skouteris 2010). Borrowing from Michelle Wright, linear progress narratives are problematic on at least two fronts, logical and empirical. Logically, the necessary point of index, or origin, presupposes a backwards-looking principle that restricts forward-looking horizons: 'On a linear timeline, origins [. . .] define the entire timeline [. . .] [and] inhibit radical revisions' (Wright 2015, 46). Empirically, 'the white Western linear progress narrative is linear only through strenuous manipulation of the facts' (Wright 2015, 40). As an example of this point, international law has been enrolled into the production of time units tailored to linear measurement in increasingly precise increments – though the material production of linear time under law is hardly linear – and in turn international law has enrolled the same time standards for its own practical propagation (Gordon 2021a). In addition to serving as a support for a progress narrative, this technological co-production also supports international law as an exercise in stabilising expectations, which includes sustaining expectations despite disappointment (Luhmann 2004). This puts mainstream international law in a dual discursive position: predicated on the possibility of frustrating hope(ful expectations) in any given case, while sustaining the relational work of hope(ful expectations) across all cases.

## A Non-Linear Temporal Frame

Against the foregoing backdrop, I would like to situate and consider the activist work associated with Black Quantum Futurism (BQF), which might qualify as what has recently been referred to in anthropological circles as time trickery, sometimes future trickery (Moroşanu and Ringel 2016; Phillips 2015; Ringel 2016). Time trickery describes an everyday ability to 'manipulate temporal processes: [to] actually slow down or speed up our own practices in relation to those of others and to alternative expectations and probabilities; [to] install specific rhythms, structures, and temporal orders that will coordinate social life in the future' (Ringel 2016, 26–7). BQF exhibits elements of time trickery. It is an 'approach to living and experiencing reality by way of the manipulation of space-time', which 'derives its facets, tenets, and qualities from quantum physics, futurist traditions, and Black/African cultural traditions of consciousness, time, and space' (Phillips 2015, 11). Some of it is really far out. Its introductory text, which I rely on here, is a zine, and includes a contribution that proposes to create a brief sound that can, when emitted, 'deactivate all money' (Stanley 2015, 58).[1] But BQF is also a grounded programme. In part, it is grounded through the activities of the figures associated with it, such as Rasheedah Phillips: lawyer, recognised public housing advocate and community activist in Philadelphia. In part, the project is grounded in the nature of its performances, and a number of those associated with the project are performing artists.

BQF announces an audacious ambition: 'to see into possible futures and/or collapse space-time into a desired future', thereby also to 'subvert the strict chronological hierarchical characteristic of linear time' (Phillips 2015, 11; 12). The audaciousness is rooted in multiple traditions of theory and praxis. Among them, one is a connection to Black prophetic tradition, which I will touch on briefly in conclusion (Marable 2012). A second is by recourse to quantum theory, which has been making tangible inroads against temporally linear constructs, in the physical and social sciences (Wright 2015).[2] A third is made clear by Phillips's collaborator, Camae Ayewa (writing as Camae Defstar), who frames the introductory zine with 'the troubling reality of memory and how it plays a role in our daily lives' (Defstar 2015, 7). The memory that Ayewa invokes is one of violence and oppression characterising political, economic, carceral and other systems that punish Black bodies, and is refreshed every day. This memory is the archive and register of (overcoming) what Denise Ferreira da Silva describes as total violence (Ferreira da Silva 2022). The ongoing reality of that memory, the trauma that it sustains and is sustained by, are not things to forget or get beyond. Rather, they are constant grounds for a tenacious vitality: 'The stigmatizing journey through a highly technological, physiological and continuous bondage has brought forth many forms of healing out of the Black community in America' (Defstar 2015, 8).

The grounding in memory, the archive of overcoming violence, and the problematisation of linear time are closely interlinked. As Michelle Wright argues, 'the site of trauma is often displaced in the act of linear return' (2015, 83). Like Wright, Ayewa and other contributors to Black Quantum Futurism envision a different, non-linear relationship to the site of trauma, in which foundational violence remains active in the present, whether in the form of suppressed futures and pasts or in the constant overcoming, or becoming despite that violence and suppression. A complex condition is manifest in this non-linear relationship to trauma. I refer to it here with a loosely psychoanalytic signifier, the barred subject, which I will expand on at points (Žižek 2008).

The barred subject is something like an inceptive force that is always at work but never accessible.[3] That combination of inceptive/always/never presents a peculiar temporal situation that defies linear representation, but is appropriate to the role of memory and trauma in the vitality that Ayewa describes. But going beyond the psychoanalytic signifier, I will refer specifically to the barred Black subject, for the denial of subjectivity across political and legal systems on the basis of racialised bodies. Recent Black social theory, some of it inspired by Frantz Fanon, has explored the barred Black subject in various ways, including Zahi Zalloua's analysis of Black being (2021), works by Stefano Harney and Fred Moten (together and individually) exploring the Black subject, and Frank Wilderson III's examination of Black subjectivity, or subjectivity under erasure (2010: xi). In each of these works, recognition of the violence imposed on the barred Black subject supports a critique of modernity and foregrounds the Black life that persists despite modernity's violence. Zalloua encapsulates the critique: 'Modernity invented blackness: "modernity gave birth to" black being "through dispossession and abjection"' (2021, 147, quoting Warren 2018, 28).

Where the sovereign subject is part of the fundament of modern international law and relations, and an engine of the violence around which the institutions of international law and relations are organised (Knox 2020), the barred Black subject marks here the persistence of life despite that fundamental violence (Hartman 2022; Wynter 2003). The barred Black subject gives the lie to the impossible condition signified by modern sovereignty, which simultaneously prescribes and frustrates subjectivity. Thus, Zalloua describes the defiant vitality of Blackness, the power of which 'lies in lack itself, delighting in the non-all of civil society, in the "structural inconsistency" of the Symbolic. Blackness discloses a world [...] where social death is not destiny, where blackness, in the mode of the parahuman, stubbornly and inventively remains' (2021, 165). Rob Knox has examined the racialised subject in international law with a combination of Fanon and stretched Marxism, to use his term, linking 'processes of racialization [. . .] [with] the logic of capital accumulation' in a complex interrelationship of race, law, value and imperialism (2016, 126). In addition, Marie Petersmann, Dimitri van den Meerssche and others have been exploring the valence of critiques founded in the barred Black subject for issues of global governance raised by practices of algorithmic violence, pointing to lessons of fugitivity, among other things (Petersmann and van den Meerssche 2023). These themes of capitalist determination, on the one hand, and fugitivity, on the other, come up again below in the context of works by Marquis Bey and Harney and Moten.

The barred subject also presents an alternative to the familiar analytic of the constitutive outside. While the analytic of the constitutive outside has done significant work to diagnose oppressive social structures, it is limited insofar as the constitutive outside remains a story about the inside that it demarcates (Wright 2015). The barred Black subject tells another story, one that defies the inside/outside binary, to describe a unique and uniquely vivacious identity of Black being bound up with abyssal trauma and resistance. I return to this below, in considerations guided by Moten and Denise Ferreira da Silva, and will try to unpack these arguments as I go. For now, the possibility that I take up here, as put by Ayewa, is that '[l]ooking through the lens of Black Quantum Futurism' yields a vision of 'hope in a dystopian reality', and the dystopian reality is not a disavowed one (Defstar 2015, 7). And in this way, BQF aligns with other expressions of hope, like Joseph Winters's *Hope Draped in Black* and Bey's 'Fugitive Hope' (Winters 2016; Bey 2018). I will now turn to these aligned expressions of hope, before contrasting BQF with a different articulation of quantum optimism, by the European Union (EU) as part of its Digital Decade.

**Hope of No Hope**

Winters, whose title borrows from Adorno, links hope with melancholy, exploring the Black literary tradition for 'conceptions of hope and futurity that are mediated by melancholy, loss, and a recalcitrant sense of tragedy' (2016, 6–7). The point is not to turn hope into sadness, but to suggest, with authors like Du Bois, Ellison and Morrison, 'that the possibility of a better world involves a heightened capacity to remember, register, and contemplate the damages, losses, and erasures of the past and present' (Winters 2016, 7). On this basis, Winters 'contends that a better world, a more generous world, involves being more receptive to those dissonant, uncomfortable dimensions of life and history that threaten our sense of stability, coherence, and achievement' (2016,

7). The opposite of Winters's melancholic hope is progress, and the distance between them is a site of struggle: 'If progress is the condition of the possibility of hope in our culture, then this is a hope that has little to no room for melancholy' (2016, 15).

Bey largely shares Winters's observations and arguments, and also makes clear the vivaciousness of a hope draped in black. Bey works against 'shallow, platitudinal notions of hope' to describe instead a tenacious hope at the foundation of 'an ongoing, quotidian praxis of defiance' (2018, 4). They call this a fugitive hope, connecting up with other work on fugitivity, gestured to above (Harney and Moten 2013). Fugitive hope mixes motion and stability. On the one hand, 'fugitive hope is the escape from fixed rootedness [. . .] which is thus a radical opening of possible futurity, which is, further, the work of that incomplete, future-oriented posture of abolition' (Bey 2018, 5). On the other hand, fugitive hope 'demands that we live here, otherwise [. . .]. We have to live right here because . . . "We ain't goin' nowhere"' (2018, 11; quoting Howard 2015). The seeming contradiction emerges out of historical praxis, becoming legible in an 'examination of how and why, by what means and through what livelihood, those purportedly condemned to enslavement, fungibility, and onticidal nonbeing continued to have the unbounded capacity for being otherwise than those circumscriptions' (Bey 2018, 5).

Here again is the barred Black subject, or Zalloua's Black ~~being~~, denied subjectivity as equal participant in modern civil society, but a fundamental participant nonetheless, and one that has persisted by sustaining life in the place of denial – by the refusal of what has been refused, as Stefano Harney and Fred Moten put it (Harney and Moten 2013, 96). The continuity between (the refusal of) civil society and the transboundary violence of the Middle Passage supports the salience of fugitivity and the barred Black subject for international legal critique, as Petersmann and van den Meerssche have begun to elaborate (2023). Going further, the examination that Bey proposes reaches, from cruel optimism's personal and affective register, to social systems on a global scale: 'I want to make explicit not only the self-transformation that occurs when working through these damages – damages that are epitomised by the historical violence done to Blackness – but also the radical *world* transformation' (2018, 5, italics in original). Those transformative possibilities are prefigured in the vitality of Black elegies, another link between Bey's work and Winters's, along with the many others who find emancipatory praxis in Black lyrical traditions: 'My understanding of Black elegy, though historically a poetic of death and mourning, is always mired in life' (Bey 2018, 7).

Moten is among those others who celebrate musical traditions in a similar register. Playing on mourning/morning, he conveys the transformation that Bey describes, here in terms of a utopia 'reconfigured in a morning song, at morning time, by a moan of pain and joy' (2017, 85). The morning song defies the death of objectification, but not by attaining to the subjectification that imposed object status on Black bodies in the first place. Here is a point that I promised to return to earlier: the barred Black subject, the Blackness that is barred (socially, carcerally and otherwise) from being a subject – or for being a subject – defies the dominant binary of subject/object. In part, the barred Black subject defies the subject/object binary through the experience of what Ferreira da Silva relates through slavery's legacy of total violence:

> A human among use values, her expressions of pain and pleasure are liable to suffer repression, and therefore she does not fully belong to the world of commodities. And as a trade object, a use value among persons, her presumed interiority is available for occupation, and therefore she does not fully belong to the world of subject of pleasure and pain. (Ferreira da Silva 2022, 281)

The persistence of fugitive life in the face of that double repression points to a mode of Black being, or Zalloua's Black ~~being~~, that does not conform to modernity's subject/object distinction and the Kantian inside/outside with which it is associated (Ferreira da Silva 2022). Thus, following Moten, the morning song defies the death of objectification not by 'an internalization [among the objectified] of the outside [occupied by subjects]', but by 'the possibility of the exteriority of the inside, the becoming-object of the speaking, singing, commodified object' (2017, 85). Moten describes the transformation in a way that makes clear the aim not to fall back into the binary system of subject/object, inside/outside, hope/melancholy, etc:

> This becoming-object of the object, this resistance of the object that is (black) performance, that is the ongoing reproduction of the black radical tradition, that is the black proletarianization of the bourgeois form [. . .] is the activation of an exteriority that is out from the outside, cutting the inside/outside circuitry of mourning and melancholia. (2017, 85, acknowledging Derrida)

Crucially, while the barred Black subject defies modernity's subject/object and inside/outside binaries, it does so from within the constraints of modern civil society. In this way, Black ~~being~~ is not some alien force, but an overcoming that participates in modernity itself. As Zalloua puts it:

> blackness compromises the [modern] human, de-completes its being – figures the inhuman – sickens and alters ontology, and infiltrates the culture of whiteness only to expose its fantasies of mastery and control, short-circuiting its framing, its racist machinery from within. (Zalloua 2021, 166)

Moten's gesture, above, to proletarianisation and bourgeois form speaks directly to modern fantasies of mastery and control, and points to another inimical notion and application of hope, alongside cruel optimism, progress and more platitudes. I turn to Moten's work with Harney on *The Undercommons* (2013) to explicate briefly, and to bring this into the register of contemporary governance and capitalist determination, as mentioned above, before moving precisely to these inimical applications of hope as they exist in the quantum programme that is part of the EU's Digital Decade.

In *The Undercommons: Fugitive Planning and Black Study*, hope is a governance technology at work in the service of contingency and precariousness. In a totalised neoliberal condition, in which market imperatives to manage and exploit contingency would encompass every aspect of living, hope is the means of enrolling a population. Harney and Moten make the point with a vocabulary of policy and planning, in which policy is the production of instability and precariousness as part of the economic valorisation of contingency and adaptation, whereas (fugitive) planning is the defiant assertion of a life force that resists the destabilising imperatives of the market. The political economic condition is clear: 'Every utterance of policy, no matter its intent or content, is first and foremost a demonstration of one's ability to be close to the top in the hierarchy of the post-fordist economy' (2013, 77). Harney and Moten caricaturise those policy utterances:

> What's wrong with them? They won't change. They won't embrace change. They've lost hope. So say the policy deputies. They [the planners, according to policy] need to be given hope. They need to see that change is the only option. By change what the policy deputies mean is contingency, risk, flexibility, and adaptability to the groundless ground of the hollow capitalist subject, in the realm of automatic subjection that is capital. (2013, 76)

Hope is a key tool because this mode of governance cannot rely on brute coercion alone. Rather, '[i]t is crucial that planners choose to participate. Policy is a mass effort' (2013, 79). For these reasons, hope is instrumental for a policy of 'participation in change [and] participation as change' (2013, 80). This cynical hope is not just a tool, but a weapon:

> This is the hope policy rolls like tear gas into the undercommons. Policy not only tries to impose this hope, but also enacts it. Those who dwell in policy do so not just by invoking contingency but by riding it, and so, in a sense, proving it. Those who dwell in policy are prepared. They are legible to change, liable to change, lendable to change. (Harney and Moten 2013, 80)

## Another Quantum Initiative

Let me summarise. On the one hand, there is hope draped in black, defiant, mired in life, and melancholic in a way that recalls the vitality of mourning songs and elegies in a Black tradition. On the other hand, there is cruel optimism, the progress narrative, and the cynical hope of market-driven governance in late capitalist society. Against this backdrop, I turn now to the European Commission's Quantum Flagship initiative (hereafter EQF, or Flagship), part of the Commission's policy leadership for a European Digital Decade, especially to contrast it with the quantum optimism of BQF. The EQF and BQF are both programmes that would build on advances in quantum sciences and 'share the wealth', so to speak. Both require and rely on popularising hard-to-understand quantum phenomena for societal adoption and intervention. They differ, however, in the imaginaries that they associate with the hard-to-understand character of quantum phenomena, and so differ in how they would build, and how they would share the wealth. Bringing them together here allows me to examine the politics of hope that their interaction prefigures in the complex environment of the Anthropocene.

The Flagship, launched in 2018, is 'one of the largest and most ambitious research initiatives of the European Union' (EQF 2023a). The Flagship will operate for at least ten years with a budget of at least a €1 billion. It promises that:

> the performance increase resulting from Quantum Technologies will yield unprecedented computing power, guarantee data privacy and communication security, and provide ultra-high precision synchronization and measurements for a range of applications available to everyone locally and in the cloud. (EQF 2023b)

To examine its future, I focus here on the EQF's Strategic Research Agenda, especially for what that document points up about the development of quantum information technologies for the European digital ecosystem. The Strategic Research Agenda begins with a premise not limited to quantum information technologies, namely, that 'the mastery of deep [digital] technologies will determine the future prosperity of countries and regions across the world' (EQF 2020, 8). Specific to quantum information technologies, use cases are identified 'in the fields of: medicine; physics; chemistry; biology; geo-physics; climate science; environmental sciences; mobility; defence, and data storage and processing' (EQF 2020, 60). Among other things, 'defence systems and autonomous mobility and navigation will profit from long-term stable rotation and acceleration sensors based on quantum technologies' (EQF 2020, 60). On these bases, documentation for the Flagship characterises quantum information technologies as 'essential building block[s] for Europe's technological sovereignty' (EQF 2020, 12). Technological (or digital) sovereignty has become a term of art to mark a transposition of classical sovereignty into 'a new global governing logic through which sovereignty operates' (Couture and Toupin 2019, 2311).

Along with this, 'a robust and secure communication infrastructure based on quantum security will be essential to protect European sovereignty and its economy in the face of increasing cybersecurity challenges' (EQF 2020, 23). The Strategic Research Agenda envisages a hypercompetitive economic terrain, in which '[t]he ability to process data fast will be a key driver for the future economy, where even marginal technological differences lead to valuable competitive advantages' (EQF 2020, 39). But as part of this vision of competitiveness, the possibility of disruption – and the possibility of exploiting disruption – is also a key dynamic:

> Quantum technologies have a huge potential for innovation that may revolutionise the information economy. Europe can play a leading role through strategic international cooperation to develop competitive collaborations [. . .]. Quantum technologies are one of the most disruptive R&D sectors as they present a gamechanger for the entire information and data value chain from sensing, to communication, sorting, simulating, predicting and computing. (EQF 2020, 91–2)

Despite the disruptive potential, however, the document describes a vision in which quantum technologies are inserted into contemporary infrastructures, to build a quantum network out of an already-existing architecture: 'The long-term vision is to develop a Europewide quantum network that complements and expands the current digital infrastructure, laying the foundations for a quantum internet' (EQF 2020, 22).

In these architectures, quantum technologies appear likely to continue a trend already apparent with platforms featuring artificial intelligence technologies: the costs of the expertise and computer power to run a scalable programme are high, and the competitive incentives to dominate are extreme, leading to a remarkably small club of global providers, who leverage their programmes (and control access) through cloud-based platform distribution (Rieder et al. 2021). Equally, the securitisation of quantum information technologies and the digital infrastructures into which they are to be inserted is ringfencing their development. The majority of international legal activity today is aimed at restricting access to the technologies, with market-oriented tools like import/export controls, dual use restrictions, and sanctions lists barring distribution of the technologies (van Daalen 2022). These tools favour a paranoid security apparatus that rewards the consolidation of powers.[4]

There are at least two things that the EQF has in common with BQF. Like with BQF, some of the claims associated with the EQF are well grounded, some are really far out. And both the EQF and BQF mix their temporal registers, leveraging the future for the present and leveraging the past for the future (BQF arguably goes a bit further, also leveraging the future for the past). But they do this to different ends. BQF, like the other examples of hope draped in black, builds its hope for the future on an unorthodox recognition of trauma, melancholy and mourning. The EQF builds its hopeful future out of hegemonic powers maintained with security and surplus value (van den Meerssche and Gordon 2023). The EQF is part of a regional drive for technological sovereignty that would reproduce the world/s associated with sovereign states and sovereign citizens (Gordon 2023). Its mode is governance, but its primary register is economic; its promise invokes disruptive contingency, but its purpose aims at continuity with established patterns of political-economic distribution – this mix approximates the cynical hope described by Harney and Moten as well as aspects of Berlant's cruel optimism.

In short, the EQF is the legal-institutional reverse of BQF's activist programme. But the opposed ends of the EQF and BQF are not simply opposites. They are entangled. Rasheedah Phillips, for instance, conjured the EQF before it was established, by pointing for support to an increasing incorporation of quantum physics in 'everyday language and lives' (2015, 16). Likewise, the EQF relies on taxpayer support, and is occupied with the uptake of quantum sciences in public discourses (QTE 2023). In addition, Harney and Moten point up from another angle that the relationship is not adversarial in a strictly binary way, insofar as the governance (policy) practices behind institutions like the EQF invite populations (planners) to participate in the reproduction of the world for which the institution stands.

**Interference Patterns**

On the basis, in part, of entanglements between the EQF and BQF, I do not take up their conflict as a straightforward one, involving simple opposition to one world in support of another. Instead, I am inclined to regard their entanglements and conflicts through the lens of what has been recently put forward under the neologism of chronocenosis, by Dan Edelstein, Stefanos Geroulanos and Natasha Wheatley. Chronocenosis, as they describe it, 'is a way of theorizing not simply the multiplicity but also the conflict of temporal regimes operating in any given moment' (Edelstein et al. 2020, 4). More than that, they maintain 'that power operates by arranging, managing, and scaling temporal regimes and conflicts' and 'chronocenosis is this complex and volatile intersection of competing temporal regimes that allows for particular ones among them to appear dominant at given points in time' (2020, 4; 9). Chronocenotic inquiry might, moreover, productively be merged with a diffractive method such as Karen Barad has developed to look for interference patterns among entangled things (Barad 2007, 71–96). In fine, diffractive method observes what emerges when a projection from one source passes through something else, like a grate. Thus, as diffractive method attends to the products of interactions, analytically applied it can assist in understanding the products of entanglements between BQF and the EQF, as opposed to merely charting their differences. This dovetails well with the description of chronocenosis as a tool 'for a conceptualization of different temporal regimes that moves beyond the description of their multiplicity to study their mutual interaction and competition' (Edelstein et al. 2020, 4). The lens of chronocenosis allows observation of 'multiple temporal regimes [. . .] [as they] inhabit a complex temporal ecosystem with intricate patterns of reliance, adaptation, and violence' (Edelstein et al. 2020, 27), while diffractive method helps to keep a focus on the consequences of their interaction for dynamics of subject and object, such as are at stake with hope draped in black, by 'investigat[ing] the material-discursive boundary-making practices that produce "objects" and "subjects" and other differences out of, and in terms of, a changing relationality' (Barad 2007, 93).

What does it mean to look for interference patterns in the interaction and conflict of multiple temporalities, typified here by BQF and the EQF, by hope draped in black and cruel optimism? If the temporality and cruel optimism of the EQF is characterised by progress, the melancholic hope of BQF might be characterised instead by prophecy. Cornel West and others have long pointed to the Black prophetic tradition as a mode of transformative struggle (West 1985), which Phillips channels when she writes of becoming 'the active agent in the synchronicity/focal point [that defines the future, past and present], instead of time being the active agent defining the synchronicity' (2015, 28). The progressive technological hope of the EQF is conservative in nature, designed to conserve sovereignty and wealth enjoyed by European states. Its purpose is reproductive, even as it is intended to anticipate and exploit (economic) disruption. This tension, between conservation and disruption, suggests the particular reproductive character of the EQF: to reproduce the market that Harney and Moten identify with policy in *The Undercommons*, mentioned above, which carries with it governance as/of a ubiquitous competition saturated with contingency. The EQF draws on contemporary technologies to project the past into the future; BQF, on the other hand, draws on melancholic hope to redeem the future in the present and past. As Bey, Winters, and so many of the authors cited earlier have made clear, however, this programme is also conservative, mired in tradition and tenacious vitality.

The hope that BQF exhibits is by far the more lively and vivacious, but for all its differences from the EQF, their realities are not independent. Neither, however, is the image that emerges out of their interaction a singular one. Their interaction will appear different from the different vantages and devices by which they are brought into contact. The devices that I have used and my vantage, though critical, are also institutional, and from this vantage, the image that arises out of the entanglement of the EQF and BQF is a complex one. The BQF demonstrates the relative paucity of the goals of the EQF, and the failure to conceive of socio-technical futures along anything other than a linear timeline which would reproduce familiar subjects, objects and a chimerical progress. But a hazard of the relationship that Harney and Moten describe between policy elites and planners bears watching: the imaginary of the EQF seeks the imaginative participation of the undercommons and its planners, such as the BQF collective. And it is the special technique of contemporary policy, as described by Harney and Moten, to adapt and thrive on disruption, such as fugitive planners might otherwise achieve (Harney and Moten 2013, 74–8). It is as though the contemporary policy institution has internalised a Foucauldian observation of power and counterpower, and makes ready to reproduce itself precisely with the tools that would counter it.

On that basis, the entanglement of the EQF and BQF potentially marks not the end of modernity and its cruel optimism, but their continuation. That conclusion, however, also assumes a linear, one-to-one relationship between power and counterpower. The concern about power–counterpower, cooption and continuation may be viewed as part of an ersatz, linear temporality, and indeed a perpetuation of its hegemonic character. In a roughly analogous discussion, Michelle Wright points out the injustice done to James Baldwin by reading his autobiographical work through a rigidly linear timeframe, reducing him to one-dimensional caricature in the process (2015, 116). Likewise, BQF may exceed my observations and the frame that comes with them. As Ferreira da Silva argues, 'in the quantic moment, the *I Think* has no capacity to claim that its institutions apprehend an already ordered world' (2022, 262, italics in the original). And quantic moment or no, BQF praxis partakes in 'a fundamental paradox of pushing through a reality while simultaneously experiencing it' (Defstar 2015, 9). From this paradox springs hope draped in black.

## Notes

1. There are a number of works produced under the BQF aegis to this point. I focus here strictly on the introductory zine for the purposes of this short chapter.
2. There has been a movement lately to adopt such a programme within the field of international relations, spearheaded by figures such as James der Derian, Alexander Wendt and Laura Zanotti (der Derian and Wendt 2020; Zanotti 2018). This chapter, however, is differently oriented to quantum imaginaries, and does not take up their transformative ambitions within its limited scope.
3. The psychoanalytic tradition that I draw on for this starting point derives from Lacan's reading of Freud, and, in turn, Žižek's reading of Lacan. Lacan's teaching is foundational throughout this chapter for his analysis of a decentred subject, which includes, in exceedingly reductive terms for summary purposes here, an excessive vitality and a constitutive Other (Lacan 2006). Lacan developed in this context the matheme of the barred $, denoting fundamental division and preclusion in the figure of the subject (Lacan 1961). The aim of this chapter, however, is not Lacanian analysis. Rather than remain with Lacanian exposition, I will elaborate and explore the barred subject by reference to contemporary invocations and examinations of the notion.
4. The last two paragraphs are shared substantially with my article on digital sovereignty and international legal imagination (Gordon 2023).

## References

Barad, K. 2007. *Meeting the Universe Halfway*. Durham, NC: Duke University Press.

Berlant, L. 2011. *Cruel Optimism*. Durham, NC: Duke University Press.

Bey, M. 2018. Fugitive hope: The constitutive life in Black elegies. *Callaloo* 41.2 (2018): 4–12.

Couture, S. and Toupin, S. 2019. What does the notion of 'sovereignty' mean when referring to the digital? *New Media & Society* 21(10): 2305–22.

Defstar, C. 2015. Forethought. In R. Phillips (ed.), *Black Quantum Futurism: Theory & Practice*. AfroFuturist Affair, House of Future Science Books, 7–10.

Der Derian, J. and Wendt, A. 2020. 'Quantizing international relations': The case for quantum approaches to international theory and security practice. *Security Dialogue* 51(5): 399–413.

Edelstein, D., Geroulanos, S. and Wheatley, N. 2020. Chronocenosis: An introduction to power and time. In D. Edelstein, S. Geroulanos and N. Wheatley (eds), *Power and Time: Temporalities in Conflict and the Making of History*. Chicago, IL: University of Chicago Press, 1–50.
EQF (European Quantum Flagship). 2020. Strategic Research Agenda. Available at https://qt.eu/app/uploads/2020/04/Strategic_Research-_Agenda_d_FINAL.pdf (accessed 3 January 2024).
EQF (European Quantum Flagship). 2023a. Website, available at https://qtedu.eu/ (accessed 3 January 2024).
EQF (European Quantum Flagship). 2023b. Introduction to the Quantum Flagship, available at https://qt.eu/about-quantum-flagship/introduction-to-the-quantum-flagship/ (accessed 3 January 2024).
Ferreira da Silva, D. 2022. *Unpayable Debt*. London: Sternberg Press.
Gordon, G. 2021a. Engaging an infrastructure of time production with international law. *London Review of International Law* 9(3): 319–49.
Gordon, G. 2021b. The time of contingency in international law. In I. Venzke and K. J. Heller (eds), *Contingency in International Law: On the Possibility of Different Legal Histories*. Oxford: Oxford University Press, 162–78.
Gordon, G. 2023. Digital sovereignty, digital infrastructures and quantum horizons. *AI & Society*, https://doi.org/10.1007/s00146-023-01729-7.
Harney, S. and Moten, F. 2013. *The Undercommons: Fugitive Planning and Black Study*. New York: Minor Compositions.
Hartman, S. 2022. *Scenes of Subjection: Terror, Slavery, and Self-making in Nineteenth-century America*. New York: W. W. Norton.
Howard, Z. *Won't Die Things*. 2015. YouTube. Available at https://www.youtube.com/watch?v=xbtxa1XJ7gA
Knox, R. 2016. Valuing race? Stretched Marxism and the logic of imperialism. *London Review of International Law* 4(1): 81–126.
Knox, R. 2020. Haiti at the league of nations: Racialisation, accumulation and representation. *Melbourne Journal of International Law* 21: 245.
Lacan, J. 1961. Seminar IX: Identification. Available at http://www.lacaninireland.com/web/wp-content/uploads/2010/06/Seminar-IX-Amended-Iby-MCL-7.NOV_.20111.pdf (accessed 3 January 2024).
Lacan, J. 2006. *Écrits*. New York: W. W. Norton.
Luhmann, N. 2004. *Law as a Social System*. Oxford: Oxford University Press.
Marable, M. 2012. The Black faith of W. E. B. Du Bois. In D. B. Chambers (ed.), *The Past Is Not Dead: Essays from the Southern Quarterly* 23: 149.
Meerssche, D. van den and Gordon, G. 2020. 'A new normative architecture': Risk and resilience as routines of un-governance. *Transnational Legal Theory* 11(3): 267–99.

Meerssche, D. van den and Gordon, G. 2023. The contemporary values of operadiction regimes. In I. Feichtner and G. Gordon (eds), *Constitutions of Value*. Abingdon: Routledge, 236–54.

Moroşanu, R. and Ringel, F. 2016. Time-tricking: A general introduction. *The Cambridge Journal of Anthropology* 34(1): 17–21.

Moten, F. 2017. *Black and Blur*. Durham, NC: Duke University Press.

Petersmann, M. and Meerssche, van den D. 2023. On phantom publics, clusters and collectives – Be(com)ing subject in algorithmic times. *AI & Society*, https://doi.org/10.1007/s00146-023-01728-8.

Phillips, R. 2015. Black quantum futurism: Theory and practice – Part One. In R. Phillips (ed.), *Black Quantum Futurism: Theory & Practice*. AfroFuturist Affair, House of Future Science Books, 11–30.

QTE (Quantum Technology Education portal). 2023. Available at https://qtedu.eu/ (accessed 3 Januaary 2024).

Rieder, B., Sileno, G. and Gordon, G. 2021. A new AI lexicon: Monopolization: Concentrated power and economic embeddings in ML & AI. In *A New AI Lexicon: Responses and Challenges to the Critical AI discourse*. AI Now Institute, 1 October. Available at https://ainowinstitute.org/publication/a-new-ai-lexicon-monopolization (accessed 16 January 2024).

Ringel, F. 2016. Can time be tricked? A theoretical introduction. *The Cambridge Journal of Anthropology* 34(1): 22–31.

Skouteris, T. 2010. *The Notion of Progress in International Law Discourse*. The Hague: TMC Asser Press.

Stanley, T. 2015. Emancipation 150: The great jubilee. In R. Phillips (ed.), *Black Quantum Futurism: Theory & Practice*. AfroFuturist Affair, House of Future Science Books, 55–60.

van Daalen, O. 2022. Making and Breaking with Science and Conscience: The Human Rights-Compatibility of Information Security Governance in the Context of Quantum Computing and Encryption. PhD dissertation, University of Amsterdam.

Venzke, I. and Heller, K. J. (eds). 2021. *Contingency in International Law: On the Possibility of Different Legal Histories*. Oxford: Oxford University Press.

Warren, C. L. 2018. *Ontological Terror: Blackness, Nihilism, and Emancipation*. Durham, NC: Duke University Press.

West, C. 1985. The prophetic tradition in Afro-America. *Drew Gateway* 55 (2/3): 97–108.

Wilderson III, F. 2010. *Red, White & Black: Cinema and the Structure of US Antagonisms*. Durham, NC: Duke University Press.

Winters, J. 2016. *Hope Draped in Black: Race, Melancholy, and the Agony of Progress*. Durham, NC: Duke University Press.

Wright, M. 2015. *Physics of Blackness: Beyond the Middle Passage Epistemology*. Minneapolis: University of Minnesota Press.

Wynter, S. 2003. Unsettling the coloniality of being/power/truth/freedom: Towards the human, after man, its overrepresentation – An argument. *CR: The New Centennial Review* 3(3): 257–337.

Zalloua, Z. 2021. *Being Posthuman: Ontologies of the Future*. London: Bloomsbury Academic.

Zanotti, L. 2018. *Ontological Entanglements, Agency and Ethics in International Relations: Exploring the Crossroads*. Abingdon: Routledge.

Žižek, S. 2008. *The Ticklish Subject: The Absent Centre of Political Ontology*. London: Verso.

# 10

# In the Breaches of Cancelled Futures: The Entropies of Modernisation and Ecological Recomposition

*Renan Porto*

## Introduction

I am Brazilian. Every day when I take my phone, open my social media and scroll down the screen, I see notices of Indigenous people being killed by gunmen hired by farmers or miners; Black people being killed or subject to other forms of aggression from police in favelas; rainforest being burned or cut off; many regions of the country – including the region where I came from – being affected by intense floods and many people losing the little they have when not their lives. I have the phone in my hand. The Internet is an overflow of information in my brain. I am a PhD researcher in London. Most of my time here is very lonely. When I am working, I also need to deal with a huge amount of information to be read, processed, evaluated, organised and presented in lectures or papers. Sometimes it takes me close to burnout. And my writing will not change the situation in Brazil. Teaching on climate change, I can see how the institutions responsible for elaborating climate policies are far from being able to change the course of global warming. My mind, my society, my planet, all of them are derailing into a possible collapse. What will bring me hope?

This chapter draws from Félix Guattari's analysis, in *The Three Ecologies* (2014), of three forms of entropy that mark the experience of the world produced by modernisation: psychic, social and environmental. Modernisation and its time, oriented towards progress and better futures, its expansion of technical interventions across all dimensions of life and over the globe, produced an experience of the world marked by mental illness, social violence and environmental collapse, mainly for non-white people in poor situations. When the future is cancelled and the present conditions are already drastically degraded, hope is not about better futures but is the affect that is enacted by actual actions and events that make possible a different way of inhabiting the present. With the inspiration of Indigenous cosmologies, this chapter points towards forms of ecological recomposition of these three interrelated dimensions of our lives – mind, society and environment – that could contribute to hope as a set of interlinked practices of ecological recomposition.

The first section approaches the psychic and social dimensions as two interconnected parts. It describes how our subjective experiences are increasingly mediated by technical devices and the impact of this on our mental health. Considering how it has become part of our ways of working, I take into consideration the situation of non-white populations that are affected not only by the mental entropy caused by the overflow of information and visual stimuli but also by racialised and militarised social contexts that are marked by a social entropy. In the second section, although not separated from the psychic and social dimensions, the focus is upon the discourses on the Anthropocene and climate change, which are key contemporary narratives on the future, mobilising hope and fear as descriptions of environmental entropy. These three forms of entropy are unfolding and unplanned consequences triggered by the new technologies of working, producing, communicating, controlling, policing and killing created by modernisation. Thus, I take the latter as the conjunction of these three entropies in the third section. In conclusion, I analyse Bruno Latour's proposal of ecologisation as an opposite vector of worldmaking in dialogue with Indigenous cosmologies. This approach enables a reworking of hope in terms of paths of ecological recomposition, able to oppose the modernisation front and create other forms of inhabiting the planet.

**Psychic and Social Entropy**

What is there to say about the future when the very conditions of the present are already degrading for millions of people and we feel it haunting and surrounding us, ever closer to the possibility of reaching a cataclysm and general collapse? How can we write about the future when we experience a growing deterioration in the quality of time lived, which is accelerating more and more, and a compression of space, which squeezes us into ever smaller and more expensive cubicles in big cities? How is it possible to write about hope when the a priori conditions of our sensibility and perception, time and space are so drastically contracted (Viveiros de Castro 2019)? The future has concrete present affects felt by the body, such as anxiety, depression and stress, besides their consequent somatisations. The body feels the time that presses upon it.

In the 1980s, Donna Haraway was elaborating a perspective of the human body as a cyborg, a cybernetic organism that couples to various technical, pharmacological, electronic and mechanical devices to expand its capacities for action and production (Haraway 1991). Four decades later, this seems even more present in contemporary life and these technical assemblies and interventions on the body fulfil heteronomous demands imposed by the need for survival and the desire to prosper in better living conditions. This body, constantly connected to these various objects, is also bombarded all the time with a flood of information and stimuli that the brain is unable to process. This addiction to visual and sound stimuli, instead of amplifying the brain's cognitive capacities, ends up making it paralysed, exhausted and frustrated, leading to a lethargic state that requires the body to rest before being subjected to this perverse cycle again.

These are not only the conditions of forms of entertainment, interaction and contemporary communication, but also of the ways of working and producing in which the digital image, subjectivity, creativity, knowledge and information are resources invested in the production of values, services, visual identities, advertising and imaginaries – something that the Italian *operaista* (workerist) tradition has called cognitive capitalism and immaterial labour (Lazzarato and Negri 2001). And depending on the colour of your skin and the geographical space where your body is located, the spatio-temporal conditions can be even more degrading when subjected to forms of necropolitical power (Mbembe 2019), going beyond contraction and reaching the point of total suppression: death. This can be caused either by direct state action through the police or by the social diffusion of violence in gangs and militias that completely pervert the liberal legal principle of the monopoly of violence by the state.

It is important to note how necropolitics is not concentrated in a sovereign's power to kill, possessed by the state, but is a form of power that configures worlds in which death circulates unrestrictedly, creating a hellish state of war among the same population. Necropolitics is also a form of fratricide (Galdino 2021). People subjected to this kind of reality have to cope not only with the excessive and infinite demand for professional qualification and production to survive in precarious jobs, but above all they have to cope with surviving the state of war to which they are exposed on a daily basis. This creates unbearable social and psychological pressure that can always spill over into new cases of violence at different levels of aggression, whether verbal or physical. In addition to inheriting the consequences of the trail of destruction left by histories of colonialism and chattel slavery, Black populations are cast to their fate in a social and political configuration in which the state mechanisms of protection and mediation of conflicts are not present and the law in force is that of necropolitics: to make die and let die (Mbembe 2019).

Thus, the experience of modernised worlds is affected by entropy either in our subjective and psychic experience or in a social dimension. These two dimensions are linked by the presence of technical devices shaping our experience of the world. On corporeal, subjective and psychic dimensions, technologies of communication, the overflow of information and audio-visual stimuli, the turning of human experience into data, and other experiences mediated technologically have been overwhelming the coping mechanisms of many. This can be described as a form of psychic entropy that alienates us from our subjectivity, imposing upon us a technical time and rhythm that can exceed mental capacities.

This is not separated from the precarisation and impoverishment of the forms of living in neoliberal societies, in which subjects cannot count on the protection and guarantees of the welfare state. For Black populations in post-colonial countries, welfare states were not even present. For them, these new forms of working under pressure without any granted future coexist with the social insecurity of militarised territories where the state can be the main perpetrator of violence. This is also intensified by military technologies brought by modernisation. Life is then surrounded by different technologies of control: either the images, sound and screens that capture our subjectivity and desire from the inside or the military apparatuses that control our bodies externally and physically. This is the world modernisation made.

What I want to show here is that the narratives on catastrophic times that usually mark contemporary discourses on the future are instead the already present for these populations. To think of hope in these contexts should be a radical engagement with the present. Hope is the affect enabled by each little action or event that opens a breach for a new tomorrow, even though the physical transformations of the planetary conditions do not respect human times and the tragedy to which these populations are cast is aggravated by extreme natural events. I will move to this topic to show it as part of the experience of the world shaped by modernisation.

**Environmental and Climatic Entropy**

The Anthropocene has been a key narrative to approach climate change and the future of human life on Earth. This can be described as a third form of entropy – environmental, climatic and cosmic – triggered by the modern and colonial forms of inhabiting and modernising the planet. If it was not enough to maintain this sickening rhythm of life in cities, the geographical form that concentrates most of the human population today, this time is becoming further compressed. Capitalism not only gives rise to a social entropy that is increasingly difficult to control by neoliberal modes of governing (Foucault 2010) – and that increasingly needs to make use of military violence to enforce the fulfilment of its economic and legal needs (Ferreira da Silva 2009) – but also a cosmic entropy (Valentim 2018), leading the entire planet and its human and nonhuman inhabitants to collapse. Considering the anthropogenic origins of the ongoing planetary transformations of climate change, Valentim coined the term *anthropy* to name the kind of cosmological and planetary disruption caused by the a*nthropos* (2018), the same prefix to qualify the geological epoch that many scientists say we have entered: the Anthropocene.

The Anthropocene is a disputed concept popularised mainly through the work of Dutch chemist Paul Crutzen, who in his famous article 'Geology of Mankind' (2002) used the concept to designate how human activity on the Earth's surface has resulted in causal proportions of a geological order. Atmospheric changes that could take thousands of years to happen in geological time have happened in the time of a human life. The Anthropocene is an epoch where humanity has become a geological agent. However, critics of the concept, such as Jason Moore (2015), argue that the concept of the Anthropocene is a generalisation that conceals the differences and inequalities between the portions of humanity most responsible for deforestation, energy consumption and carbon dioxide emissions into the atmosphere. Therefore, Moore suggests the concept of the *capitalocene* as more appropriate, arguing that it is not humanity in general that caused climate change, but a certain historical configuration of human life that is capitalism and whose command centre is in the rich countries of the Global North.

Malcom Ferdinand (2019) employs the term *negrocene* to designate the intimate relationship between environmental destruction and slavery. For Ferdinand, Western modernity is marked by both environmental and racial fractures that work in conjunction. The environmental destruction of colonised territories is a continuity of a colonial mode of inhabiting the land: exhausting its resources, extinguishing the ways of life of Indigenous peoples inhabiting it and employing enslaved labour of Black bodies to carry out the colonial enterprise. It shapes the modes of production characteristic of colonial plantation systems. In a world fractured by racism, the extreme consequences of the exhaustion of the planet, such as extreme natural events, end up aggravating already existing social inequalities. The populations most vulnerable because of the world shaped by colonial history are also the most vulnerable to climate change. Moreover, before tragedies caused by this same colonial mode of land habitation, a selective and exclusionary politics is put in place to protect the most privileged human parcels and abandon the non-white populations to their own fate even after expropriating them of their territories, their bodies and their wealth. To name this, Ferdinand used the metaphor of Noah's ark. The *negrocene* expands to a planetary scale a mode of inhuman and perverse treatment that was applied to enslaved Black bodies, sucking every last drop of blood from the strength and energy that these bodies could offer.

Donna Haraway (2016), in turn, affirms the variety of names to designate such an era that involves so many layers and proposes the term *chthulucene* to activate a tentacular imaginary that evokes the multispecies relations that make up the diverse ways of life that are made in *sympoietic* coevolution. With the concept of *sympoiesis*, Haraway conceptualises how life forms coevolve in interaction with diverse organisms and not as closed systems that self-produce as in the conception of autopoietic systems. Furthermore, Haraway says that the concept of the Anthropocene has a dubious effect by putting too much focus on the human and placing us as the main agents of geological transformation, thus invisibilising our implication with other living species that are also producing history. Naming the catastrophic agency of the human would also produce the effect of making us feel powerful enough to modify the planet with miraculous geoengineering projects. Therefore, the solutions to the mess caused by a certain part of humanity would not be so far out of reach for this same portion with privileged access to the most advanced technoscientific means. The solutions would come from the same creators of the modern fable of total technoscientific control over nature and its consequent technical and economic imposition over all regions of the planet; that portions of the human species are so narcissistic that they tend to confuse themselves with the universal: white Europeans or Europeanised whites, especially the rich, no matter what continent they are in.

Deborah Danowski and Eduardo Viveiros de Castro, in *The Ends of the World* (2016), develop a critique of the not-so-veiled ethnocentrism of proposals to recycle the modern Western tradition that has nothing more than technology and science to give them a sense of future. Their book explores different articulations of the relationship between humanity and the world. They discuss different perspectives in which humanity finds itself able to transcend the physical limitations imposed by an inert world amenable to technoscientific manipulation. Here, a humanity without a world leads paradoxically to a world without people, ultimately empty and dead. That is the experience of the Moderns. On the other hand, there are those people for whom the world is inseparable from the diverse agencies that populate and constitute it as always being made of relations, such as in Indigenous worlds. In the first case, a modern delirium in which human reason can transcend the physical barriers of the earthly environment. In the second, a humanity immanent to the world and its innumerable nonhuman agencies. The authors take seriously what Indigenous peoples have to say about the end of the world, which for them is not a future condition but a tragedy in progress since the beginning of the colonial invasions in which their worlds were invaded by aliens, those who come from outside. In this context, the most probable future is not the diffusion of the dreams of Californian ideology from Silicon Valley to the rest of the world, but the increasing precariousness and exhaustion of the conditions of existence that first arrived for Black and Indigenous peoples and will also arrive in the large capitalist centres. Confronted with this, Indigenous practices and knowledge become examples for imagining futures after the catastrophe that has already happened and is in progress: the progressive disarticulation of the spacetime articulations that make up the world and from which humanity cannot separate itself. In other words, the end of the world.

A very pertinent formulation of an Indigenous perspective on the end of the world is that of Davi Kopenawa in his book in collaboration with anthropologist Bruce Albert, *The Falling Sky* (2013). The title itself names a mythical discourse on the disintegration of the cosmic arrangement that sustains existence and prevents the sky from crashing down on us, leading the Earth back into chaos. The Yanomami concept to designate what we call the sky is *hutukara*, but in Yanomami language it is more than the sky as an atmospheric and physical dimension. The *hutukara* is something that covers and envelops our existence like the mantle of a mother who accommodates her offspring who rest and breathe under her bosom. Also, the *hutukara* cannot be separated from the *urihi a* that is her hair (Gomes and Kopenawa 2016). *Urihi a* can be translated as land-forest or world-forest, in which the forest is not a set of entities external to our existence but the very composition of the world in which we are immersed. *Urihi a* is a cosmic arrangement composed of different living species, organic and non-organic beings, such as rivers, minerals, rocks and air without which organic forms cannot subsist (Albert and Kopenawa 2013). The destruction of the *urihi a* is not the mere destruction of an environmental element that can be ignored or disregarded given the possibility of existing in any other environment or even existing without the forest. The destruction of the *urihi a* is the destruction of the world(-forest) and the matter of which existence itself is made. Since the *urihi a* and the *hutukara* make up this cosmic arrangement that encompasses us and involves us in relationships with other living species and inorganic elements – an arrangement that shamans are responsible for protecting and constantly preventing from being derailed into chaos – the fall of the sky is the name of a cosmic entropy that undoes the ecological webs that sustain life.

## Modernisation as the Conjunction of Psychic, Social and Environmental Entropies

Milton Santos, a geographer who studied the processes of urbanisation in so-called Third World countries, described modernisation as a process of imposing hegemonic technical systems on so-called underdeveloped countries, producing in these countries what he called 'derived spaces' (Machado 2019; Santos 2021). The spatial transformation caused by Western modernisation and globalisation enabled the geographical means of non-Western countries – including their techniques, their cultural forms, their traditional ways of living and producing – to be increasingly replaced by technical means that express the hegemonic mode of production, in this case, capitalism. This converted these derived spaces into distorted images of the hegemonic spaces (Santos 2021). This process also produced divided spaces in which the modern and the archaic are not completely separated from each other, but cross and blur together, both being affected by modernisation.

These processes of planetary modernisation and capitalist globalisation, that unfold in a clash with other temporalities and geographical configurations, end up causing a vectorial reversal of globalisation. If, on the one hand, these processes, which stem from the great centres of capitalism, have not been able to completely homogenise the different spaces where they have intervened, producing no more than spaces that are violently and racially segregated, on the other hand they have brought about a reverse globalisation that now falls upon their fortresses. The technical and predatory expansion of colonial capitalism over the planet is also producing what Mbembe has called 'the black becoming of the world' (Mbembe 2017). The degrading conditions that capitalism has imposed on Black bodies are also expanding globally and progressively under neoliberalism. We increasingly work hard to secure the little we have without much expectation of upward social mobility. We become more and more indebted, working with flexible bonds that ensure, each time, fewer guarantees and protections under a state that dismantles more and more of its public services. With rates of mental illness also on the rise, we continue to become more and more anxious, depressed and pharmacologised, struggling to reach a level of security and stability that economic capitalism and liberal politics can no longer provide, even for the white peoples of the countries where these technologies of government were created, much less for those who were looted and enslaved by these same economic and political forms.

The modern project, guided by rationalisation and disenchantment of the world with the aim of ordering it, produces more and more chaos, disorder and entropy, especially for those on whom modernisation was imposed by military force and who today live in a social disorder that is increasingly less governable by the state, and where the state has never really had the monopoly of violence. And if some would argue that all this is the effect of a specifically capitalist modernisation, let us not forget the other monstrous dimension of modern entropy. The vertical imposition of modernising technical systems not only clashes with the traditional *horizontalities* of non-modern peoples, but also disturbs the ecosystems and nonhuman forms of life of the territories devastated by the extraction of the physical materials of which every technical device is made. If these technical systems continue to operate and expand over the planet under socialist command, this will not save us from the environmental and climatic consequences they entail. As Yuk Hui writes (see 2016; 2021), technologies are not neutral and universal. If they are geared towards the complete domination of humans over nature, they will not lead us to a different future from the one that capitalism is producing.

As argued by Latour (2018), the Moderns ignore the fact that the planet upon which they act is not merely a passive recipient of their actions but reacts to them in unpredictable ways. Modernisation projects intervene on environments that are populated by other life forms that are active and that are part of the chemical, biological and ecological composition of such environments. Insisting on modernising the planet, submitting it to the technoscientific and rationalised domination of the human, these actions intensify the process of environmental and climatic entropy in which the planet already finds itself. Valentim argues that the modern project that excludes all agency from nature and nonhuman ways of life, in the production of history and the world, has as its reverse effect the monstrous re-intrusion of nature on a different scale, the 'intrusion of Gaia' (Stengers 2015) – no longer the nature that the human can domesticate, but the super-nature of catastrophe.

It is impressive that the indomitable multiplicity of countless forms of life that make up the pluriverses of non-modern peoples – those considered by the Moderns to be lacking in reason and history – were never a cause of cosmic entropy. Yet the 'life-world' constituted by the project of globalising modernisation has been disastrous for countless modes of being. It seems that the more the forms of life intensify and diversify, the more the Earth breathes, vibrates, colours itself and recomposes itself; while the physical and ontological elimination of all other nonhuman – or worse, non-white-male-heterosexual – forms of life is what causes the Earth's degradation. Here is a curious paradox: it is the intensive multiplication of agencies and life forms that can decrease cosmic entropy, not their reduction to the mental and economic monocultures of those who feel able to say to whom reason really belongs.

## Ecological Recomposition

This recomposition is inspired by the three inseparable ecological dimensions described by Félix Guattari (2014): psychic, social and environmental. In this chapter, Guattari's three ecologies are reworked as three forms of entropy – which also happen together – to then think about the possibility of ecological recomposition of these dimensions. A body is always in relation to a society that is always enveloped by an environment. These three dimensions are transversal to each other and shape each other. In the village of Florestal – a district of the city of Jequié in the state of Bahia, northeast Brazil – where I grew up, we have an environment surrounded by cocoa planted in the traditional *cabruca* system, which is a cultivation technique in which the cocoa is shaded by taller trees. I grew up helping my father with cocoa (Figure 10.1). I have childhood friends who still today are working with cocoa.

This is an activity commonly seen as merely manual and regarded with much prejudice in a world whose hegemonic centre of perspective is that of modernised places. Technique also functions as a vector for hierarchisation of human experiences and different social forms. However, the practice with cocoa also involves the intellectual ability and ecological knowledge of the workers who manage the crops, without which the crops will not prosper. It is necessary to have an accurate and trained perception to enter a cocoa crop and identify the demands that the trees express in the colours of their leaves and fruits, in the shapes in which their branches grow; to perceive the distribution of other plant species and how they condition the cocoa, in addition to the insects, snakes and birds that contribute or not to the production of the crops. On this point, cocoa farming ended up reducing the diversity of fauna, but the practices of landless workers' movements and Indigenous communities with agroecology seek to remedy this problem. Cocoa workers are also multispecies intellectuals who have important skills to contribute to the invention of collective alternatives of multispecies coexistence.

Authors such as Latour and Viveiros de Castro propose that hope, conceived through the possibility for an alternative to this front of incessant modernisation and extensive development – which imposes on us a frantic pace and pressure to produce, with much pain along the way that will ultimately lead us to catastrophe – lies in a cosmological reinvolvement with the Earth (Latour 2018; Viveiros de Castro 2011). As Latour says, the path is inwards into the Earth and through a politics in composition with other terrestrial beings, recognising the relations of co-dependency with them that condition our freedom. How is it possible to compose hope from the limits of scarcity that the capitalist depredation of the planet has already inflicted on us?

I see in the work of the artist, teacher, researcher and Indigenous leader Glicéria Tupinambá a great example of this. Glicéria proposed to recompose a Tupinambá mantle based on another one that she saw in the Musée du Quai Branly, in Paris. She discovered that there were several mantles like these that were stolen by European colonisers and that can be found in Europe today. For example, there is also a seventeenth-century mantle in the Royal Museum

**Figure 10.1** Father. Photo by Renan Porto.

of Arts and History in Brussels. However, as Glicéria explains in her account (Tupinambá 2020), the mantle could not be made of just any material. It should express the territory from which the mantle emerges and with which it is related. But the birds that existed 600 years ago and that provided the feathers used in the composition of the cloaks are no longer present in the region of southern Bahia where the Tupinambá people live – mainly because of the environmental destruction promoted by colonisation and capitalist expansion on Indigenous lands. Unlike the work of an engineer who has a previous project and orders the material to carry it out, Glicéria had to act as a bricoleur, working from the limited materials available and creating from the possible combinations among them.

Glicéria's recreation of the Tupinambá mantle is also an example of what Yuk Hui called 'cosmotechnics' (Hui et al. 2021). Her technique of composition derives, above all, from a relationship with her territory. The way she describes how she obtained the feathers to compose the mantle demonstrates how the material is not merely available to human interest but is a donation of what the territory with its diverse forms of life can offer (Figure 10.2). Her artistic practice is not separated from her struggle to recompose her territory and a diplomatic negotiation with other forms of life that inhabit it. Her account of the mantle is linked to her history of struggle for the retaking of ancestral lands with the Tupinambá people (Hui et al. 2021).

The paintings of Jaider Esbell, an artist of the Macuxi people, are also a strong expression of a universe in which beings always exist in relation. His works echo something similar to the concept of *urihi a*, the world-forest of the Yanomami that Kopenawa describes so well (Albert and Kopenawa 2013). As explained earlier, *urihi a* refers to the forest as a world that surrounds us, is made up of diverse life forms and populated by spirits. Esbell painted pictures that are like galaxies or meshes composed of diverse beings that intertwine with each other forming other beings that encompass them. The parts do not disappear in the whole, they do not lose their own detail, but the collective body that they form also gains its own contour. We see this, for example, in the painting *Curandeiro Trabalhando Com Tabaco* (Healer Working with Tobacco), 2020, in which the panoramic image of the painting shows us the healer smoking his pipe wrapped in the night. He is propped up by what looks like a bush if we capture the painting as a whole. But moving our gaze through the details, we see a composition of birds, frogs, wolves, people, rodents, peacocks and plants. Looking at his canvases is like entering a forest populated by spirits, where the visible beings of the trees and animals hide a spiritual double that only shamans can see, but that still gives us the feeling that we are not alone. These canvases are like a cosmic cartography of the agencies that make up the Indigenous worlds.

**Figure 10.2** Tupinambá mantle. Created by Glicéria Tupinambá.

When Kopenawa describes what he understands by ecology, he begins to list a set of beings and elements that coexist in relation, starting with the Yanomami people themselves: 'In the forest, we human beings are the "ecology." But it is equally the xapiri, the game, the trees, the rivers . . .' (Albert and Kopenawa 2013, 393). It is indeed interesting that in describing ecology by saying that it is made of more than humans — humans and, as much as them, the spiritual and nonhuman beings of the forest — Kopenawa shows how his concept of ecology is not distinct from a sociology. To describe the Yanomami community is to describe also the nonhuman beings with whom it coexists in co-dependency. Every human society is also an ecosystem of nonhuman life forms, no matter how invisibilised they may be. Bringing these agencies within the reach of our sight, hearing and touch is not separate from what they can sensorially cause us, be it delight or astonishment. Since this happens within a community that collectively welcomes this experience of composing together with the Earth, this recomposition will be corporeal, collective and planetary, taking place on the three levels considered here.

## Conclusion

Modernisation is a process that still feeds the hope of a world that can be fixed by human reason and its very calculated and planned projects. This hope takes the forms of techno-solutionism, geoengineering, accelerationism, among other contemporary discourses. What I have tried to show here is that we already live in a world made by this form of mentality. As a Brazilian, I am from a country where the promises of modernisation and development are what underpin the hope of better future. This discourse has mobilised our political imagination for more than a century already, yet what we have had until now is a country where Black and Indigenous people are killed because they do not fit in with modern projects. We are always a step back from modernisation, which is never fully realised. I tried to argue above that the world produced by such dreams of modernisation is one where we are alienated from ourselves, from other people and from other forms of life. It is a world of loneliness, isolation, mental illness, unbridled consumption, of stimulation — either chemical or audio-visual — to cover over the void of life. It is a world in which we lose the pleasure of being alive and of being involved with life. It is a way of being that steals from us the world after degrading it with its machinery, putting the human species at risk of extinction. My conclusion is that we should not feed hopes of better futures when we are not able to experience a better present. Hope is more real when it is an expression of the contentment of being what we are in the present. What Indigenous cosmologies can teach us is that life cannot be reduced to the merely productive or useful, as Ailton Krenak says (2023). Life should not be instrumentalised into projects that erase its shine. Better than development is to be involved with our realities. The concept of *urihi a* — the world-forest — is one in which we are embraced by the other forms of life that surround us. Life is a cosmic dance and I hope we can find its step again.

# References

Albert, B. and Kopenawa, D. 2013. *The Falling Sky: Words of a Yanomami Shaman.* Cambridge, MA: Harvard University Press.

Crutzen, P. 2002. Geology of mankind. *Nature* 415: 23.

Danowski, D. and Viveiros de Castro, E. 2016. *The Ends of the World.* Cambridge: Polity Press.

Ferdinand, M. 2019. *Decolonial Ecology: Thinking from the Caribbean World.* Cambridge: Polity Press.

Ferreira da Silva, D. 2009. No-bodies: Law, raciality and violence. *Griffith Law Review* 18(2): 212–36.

Foucault, M. 2010. *The Birth of Biopolitics: Lectures at the Collège de France, 1978–1979.* Picador USA.

Galdino, V. 2021. O espectro de Abel / O círculo infernal da necropolítica. *Thaumazein* 14(27): 95–109.

Gomes, A. M. R. and Kopenawa, D. 2016. O cosmo segundo os Yanomami: Hutukara e urihi. *Revista da Universidade Federal de Minas Gerais* 22(1.2): 142–59.

Guattari, F. 2014. *The Three Ecologies.* London: Bloomsbury.

Haraway, D. 1991. A cyborg manifesto: Science, technology, and socialist feminism in the late twentieth century. In *Simians, Cyborgs, and Women: The Reinvention of Nature.* New York: Routledge, 149–81.

Haraway, D. J. 2016. *Staying with the Trouble: Making Kin in the Chthulucene.* Durham, NC: Duke University Press.

Hui, Y. 2016. *The Question Concerning Technology in China: An Essay in Cosmotechnics.* London: Urbanomic.

Hui, Y. 2021. On the persistence of the non-modern. *Afterall: A Journal of Art, Context and Inquiry* 51: 60–73.

Hui, Y., Olmos, A. M. G. and Villafuerte, H. E. 2021. ¿Cosmotécnica latinoamericana? Una conversación con Yuk Hui Parte I: Antinomia de lo universal. *Technophany: A Journal for Philosophy and Technology* 1.

Krenak, A. 2023. *Life is Not Helpful.* Cambridge: Polity Press.

Latour, B. 2018. *Down to Earth: Politics in the New Climate Regime.* Cambridge: Polity Press.

Lazzarato, M. and Negri, A. 2001. *Trabalho Imaterial: formas de vida e produção de subjetividade.* DP&A Editora.

Machado, T. A. 2019. Milton Santos, Intérprete do Brasil: Uma Contribuição Geográfica ao Estudo da Realidade Brasileira. Doctoral thesis, Federal Fluminense University. Available at https://app.uff.br/riuff/handle/1/28306 (accessed 4 January 2024).

Mbembe, A. 2017. *Critique of Black Reason*. Durham, NC: Duke University Press.
Mbembe, A. 2019. *Necropolitics*. Durham, NC: Duke University Press.
Moore, J. 2015. *Capitalism in the Web of Life: Ecology and the Accumulation of Capital*. London: Verso.
Santos, M. 2021. *The Nature of Space*. Durham, NC: Duke University Press.
Stengers, I. 2015. *In Catastrophic Times: Resisting the Coming Barbarism*. Open Humanity Press.
Tupinambá, G. 2020. Curar o mundo: sobre como um manto tupinambá voltou a viver no Brasil. *n-1 edições*. Available at https://www.n-1edicoes.org/curar-o-mundo-sobre-como-um-manto-tupinamba-voltou-a-viver-no-brasil.
Valentim, M. A. 2018. *Extramundanidade e Sobrenatureza: Ensaios de Ontologia Infundamental*. Cultura e Barbárie.
Viveiros de Castro, E. 2011. Desenvolvimento econômico e reenvolvimento cosmopolítica: Da necessidade extensiva à suficiência intensiva. *Sopro: panfleto político-cultural* 51.
Viveiros de Castro, E. 2019. On models and examples: Engineers and bricoleurs in the Anthropocene. *Current Anthropology* 60(20): 296–308.

# Part III
# Negation

# 11

# Hope and the End of Critique? Crisis and Affirmation in the Anthropocene

*Valerie Waldow*

**Introduction**

Although it has often been neglected as an analytical category, hope's place in the tradition of critical thought can be traced back to the birth of critique in the eighteenth century. Reflecting on why it is that we can reasonably hope for the success of moral practice, Kant argued that we have a right to hope for better times not because things are necessarily going to be better, but because they might. Following Kant's understanding, hope is not an entity that exists outside or external to the human lifeworld but is integral to the power of reason to relate to questions which cannot be answered by experience (Kant 1998, A805; B833). This hope refers to the attitude of deeming the success of a moral undertaking possible, despite the absence of supporting evidence or probability. Kant's foundation of counterfactual hope has been reformulated by, among others, Adorno, emphasising the necessity of hope for bridging the gulf between normativity and reality, while also considering criticism of religion and metaphysics and the experience of the Holocaust. Adorno acknowledges that history has demonstrated that critique and reason do not always result in the development of rational practices that prevent self-deception. Given this inevitable element of contingency, hope now means hoping that the worst will not occur, that history will not repeat itself, that critique will fall on fertile ground (Adorno 1979, 385; 1997, 31). Hope and critique are related.

Recently, much effort has been put in to reassessing the merits of critique by academics and the political left. Challenged by the question of who is to blame for the socio-political shift to post-truth populism, authoritarianism or anti-intellectualism, there is increasing suspicion that critical approaches have contributed to these anti-emancipatory developments (Shore 2017). Although critique, at least as a buzzword, seems to be ubiquitous, there is increasing acceptance of Bruno Latour's claim that critique 'has run out of steam' (Latour 2004). It is, in particular, critical theory's approach to critique which is fundamentally challenged today. This challenge is not only due to experiences of powerlessness and insecurity in economic, political or ecological conditions, giving rise to right-wing mobilisation and identity politics, or the ongoing injustice and violence inherent to these conditions on a global scale. Critique is also being challenged from within. As criticism of the rationalism, universalism and dualism of modernist ideals intensifies, a profound rejection of negative dialectical critique and emancipatory thought has emerged, due to their association with Eurocentrism and the notion of modernity. The outcome of this trend is a broad-based interrogation of the value of critique itself.

This attack, specifically formulated with a broadly relational ontology and post-epistemological methodology in its core, is expressed under labels such as the 'critique of critique' or the 'end of critique' (Austin 2022; Bargués-Pedreny et al. 2015; Latour 2004; Thiele 2015). The matter has been amplified in debates surrounding the Anthropocene. The thesis of the end of nature, which suggests that the Anthropocene reveals a nature–culture continuum, where human history cannot be viewed as separate from geological history, is frequently associated with the notion that critique has come to an end (Bunz et al. 2017, 8, 15; Colebrook 2014, 197–8; Morton 2013, 181; Stengers 2015, 107–15). Critique, in its 'traditional' form, is thought to be not only ineffective in addressing the challenges of the Anthropocene, but also a part of the problem itself, due to its human-centred approach to the world (Colebrook 2014, 71, 198).

While critique is both ubiquitous and fundamentally challenged, utopian thinking and engagement with hope have witnessed a revival. Facing the choice between returning to a more strident ideological critique or the abandonment of any critique at all, critics increasingly call for (post)human forms of politics and new political imaginaries (Cornell and Seely 2016; Haraway 2008; Stengers 2011). However, if critique undermines its own reflexive conditions, how is it possible to speculate about a different or even better society to the given one? Hope, as introduced above, as a dimension of the capability for critical reason, does not seem to have a place in a world after critique. However, if there is no possibility for critique, what are the grounds for hope?

This chapter engages with this question by exploring the relation between hope and critique, specifically how the recent revival of hope intersects with claims that critique has ended. It discusses a corresponding change in how critique and hope are redefined within this emerging new paradigm of the Anthropocene. In doing so, it highlights a departure from traditional, future-oriented *utopian hope*, which is associated with modernist notions of emancipation and progress, to *radical hope* – a profound willingness to embrace entirely new possibilities in the face of heightened ontological vulnerability. This shift towards radical hope is closely linked to a move from critique to affirmation, which is primarily registered in post-critical conceptualisations of the challenge of the Anthropocene. This challenge urges a complete re-evaluation of human–nature relations, agency and politics, and calls for the cultivation of entirely new ways of imagining the world (Haraway 2015; 2016; Latour 2014; Tsing 2015, 2). However, as will be demonstrated, this paradigm also limits opportunities for critique. Against this backdrop, this chapter discusses the implications of the coincidence between the revival of hope and the end of critique, shedding light on both the inhibiting nature of hope as it emerges in the Anthropocene and the ongoing indispensability of critique as a vital endeavour. Ultimately, the chapter argues that hope is necessary for critical intervention in a world that still requires it, a hope that goes beyond mere wishful thinking or the affirmation of existing possibilities.

## Critique and Hope – Why Now?

Currently, it appears that where the end of critique occurs, hope emerges, suggesting that the revival of hope and the demise of critique are intertwined phenomena: hope comes into play when 'our belief, faith and trust in political or individual actions are increasingly being threatened, leading to despair and uncertainty' (Zournazi 2002, 14). Moreover, hope establishes the foundation for the potentiality of disillusionment, and it is within the confines of this disillusionment that the presence of hope manifests itself: 'Hope is always hope against the evidence [. . . that] there is a kind of discontinuity in time, there is a break, and something starts out of nowhere' (Zournazi 2002, 23). This break, this discontinuity, is closely linked to crisis, a moment of upheaval that has the power to fundamentally alter social relations, historical or political ideologies or cognitive frameworks. It marks a turning point where everything changes, despite its unpredictability (Koselleck 1982, 637–64; see also Zebrowski in Chapter 2). In these moments, in the experience of transfiguration, crisis itself might be the spark that opens up history and alternative ways of comprehending it. Here is also the place for a spark of hope.

Hope is a paradoxical concept that embodies a unique structural feature. It exists within the realm of reality, yet simultaneously serves as its counterpoint. Therefore, responding to the assumed futility of critique, a contrary perspective is stressed here: that it is the essence of critique to always entail a utopian element that is akin to the concept of hope, expressing the desire for 'a better way of living' (Levitas 1990). Without critique, in terms of analysing and dismissing the fatal dislocations of modernity, hope remains elusive. And without hope, in terms of giving reasons to engage with those failings, critique becomes meaningless. It seems therefore necessary to adopt a stance that encompasses both hope and critique. On the subjective level, hope reflects our reasons to critically engage with the world. It is not a hope that necessarily relates to belief or expectation in a better future but is crucial to prevent disappointment from turning into mere cynicism.

Problematising hope in political and moral engagement means that its 'success' is dependent on practice. Such hope would be characterised, first, by negation (of something that should not be); second, by a counterfactual element (the expectation against better knowledge or hope against reason); as well as, third, by providing an epistemic middle position between ungrounded wishing and knowledge, thus limiting the counterfactual element to hold up that hope still needs justification (Wesche 2012, 50). As a distinct mediator between *the ways things are* and *the ways things may be*, hope then matters as a catalyst for political action, but also as a link for scholars' critical engagement with the world. As such, hope provides space to be both critical and disillusioned, without abandoning the pursuit of critique as entirely meaningless. This resonates with current calls for new political imaginaries, but it also allows critical examination of the current revival of hope and its alignment with these aspirations.

## Latour's Provocation and the End of Critique

Looking at the current state of social critique presents a contradictory image. On one hand, the word 'critique' is abundantly present in various academic settings: from articles to workshops to conference themes, it seems to have become commonplace. On the other hand, critique faces a significant challenge, which registers as the end of critique, following 'Latour's provocation' (Thiele 2015, 143). In his much-cited article 'Why has critique run out of steam', Latour asks about the state of critique in a time characterised by the end of history and the Cold War divide of geopolitical ideologies (Latour 2004). Latour laments that despite becoming popular, the critical turn in social sciences to (de)construction did not translate into practical strategy. Specifically, he argues against a 'debunking impetus' of critique, deriving from the Enlightenment (Latour 2004, 323). This impetus is a manifestation of a reductionist imaginary of modernist distinctions as 'matters of facts', because it assumes a privileged understanding of reality beyond what is apparent. In response, Latour calls for the 'cultivation of a stubbornly realist attitude', an involvement with facts as 'matters of concern' which would allow for an adequate consideration of their complex constitutive processes, including their socio-political as well as their ecological, economical and ethical dimensions (2004, 231): 'The practical problem we face, if we try to go that new route, is to associate the word *criticism* with a whole set of new positive metaphors, gestures, attitudes, kneejerk reactions, habits of thoughts' (2004, 247).

One consequence of these claims is empiricism, a 'painstaking attention to traces, objects and devices' (Toscano 2012, 17). Critique is no longer about revealing ideologies and explanatory-diagnostic approaches to socio-political currents, but about the unpredictability and 'fragility of things' (Connolly 2013). This also entails a post-epistemological approach to knowledge understood to be situated, approaching the world in its immanence and emergence, 'because knowledge can no longer be extracted from its concrete context of interaction in time and space' (Chandler 2016, 406), as well as acknowledging relationality as its basic premise (Latour 2005). It also serves to radically reconceptualise political subjectivity and human agency as 'the vitality of actants' (Connolly 2011, 25), located within assemblages including nonhumans and humans alike (Latour 2005, 71).[1]

Latour's provocation has inspired a whole range of approaches to the study of world politics, such as actor-network theory, affect theory and new materialism (Bennett 2010; Cudworth and Hobden 2011). However, a growing number of commentators indicate that the radicalism claimed within these approaches reinforces the status quo rather than enabling emancipatory change, particularly by 'debunking "the human" and "human agency" at a time when neocolonial and neoliberal capitalism have perhaps never been more destructive to the vast majority of the world's inhabitants (human and non)' (Cornell and Seely 2016, 3). Koddenbrock stresses that, in focusing on the contingency and instability of things, 'any conception of unity, totality and thus, capitalism as such becomes irrelevant for intervention critique and critique in general. This diffuses and localizes the levers for transformation' (2015, 249). This focus on 'the "what is" of the world in its complex and plural emergence', as apparent in current real-time-management approaches and other adaptive forms of governance (Chandler 2016, 405), has implications for a more general understanding of responsibility and legitimation. Normative questions are handed over to the world as it is, pointing towards 'a new critical theory, which does not see the normative horizon itself but takes it from empirical analyses' (Beck 2015, cit. in Chandler 2016, 409).

While acknowledging the urgent need to critically reconsider rationalist and linear approaches of universal modernity, it is also necessary to point out that these frameworks of critique raise both theoretical and practical questions. Chandler warns:

> Today, it seems that the world [...] is emancipating humanity from its modernist aspirations. However, there is a real danger that this 'emancipation' will be even more problematic than those that have gone before.

> The posthuman world of contingency sets no 'normative horizons' beyond obedience to the external appearances of the world: the necessity of continuous adaption to the world in its emergence (Chandler 2016, 410).

In this regard, the end of critique can evoke deep pessimism about the alterability of the world, a pessimism which concerns all those who do not want to live or be governed in this way. Relation or relationality as a fundamental ontological position aims at surpassing the boundaries of human-centric thought by recognising the intricate interconnections that exist beyond our species. Illuminating the vast array of actors, both human and nonhuman, that populate our world may indeed hold potential to challenge human hubris (for a discussion, see Torrent in Chapter 14). However, without any 'unifying principles hierarchies of agency, or determinate chains of causality – any alleged criticism of what exists would [...] only reinforce the essentialised understandings of liberal modernity' (Chandler 2015, 53).

The Anthropocene marks an aggravation of this problematic. It invokes a new catastrophic scenario, a narrative of a world in constantly unfolding crisis in which humans lack the means to control the effects of their actions on the planet. In fact, it seems that it was precisely the hubris of governance as a practice of ordering and control that was actually the problem itself. For many scholars, the Anthropocene presents a significant challenge to the liberal or modernist understanding of humans and their relation to the world (Latour 2017; Morton 2013; Stengers 2015; Tsing 2015). It is hence affirmed as a historical opportunity to reach beyond established frameworks of meaning. As a particular marker of a fundamental rupture and break with modernity, the Anthropocene necessitates the break from the critical tradition (Colebrook 2014, 198). In order to embrace a new ontology, it is necessary to move beyond the project of critique. This entails rejecting traditional critical questions and instead adopting a more affirmative approach to the construction and composition of assemblages with objects (for an extensive critique, see Morgan 2019, 22).

This 'affirmative turn' in Anthropocene scholarship has evoked criticism (Bargués-Pedreny 2019; Pugh 2020; Schmidt and Koddenbrock 2019), be it for reducing humans to an adaptive role instead of challenging the status quo (Chandler 2018), or for lacking the disruptive or subversive impact it claims to invoke (Alt 2019, 140). Crucially, the Anthropocene follows a consequential logic, providing both the diagnosis and the answer: 'there *is* a nature–history continuum and [...] *therefore*, thought and agency must be construed and enabled in such a way as either to control or affirm it' (Hofstätter 2019, 8). Without a framework for critique, however, it is impossible to address these implications as ideological. Does the Anthropocene thereby suggest that in a world after critique, it is necessary to acknowledge that everything will ultimately continue exactly as it is, that 'any hope for a different future [. . .] any idea of transforming the world is humanist delirium' (Cornell and Seely 2016, 12)?

Just to be clear, there are compelling reasons to consider a modernist understanding of critique as problematic: it is dualistic, divides subject from object in a way that agency is seen only on the side of the subject and is therefore always anthropologically framed. Nevertheless, it is questionable why critical faculties should be entirely dismissed: neither the fetishisation of critique nor its replacement with affirmation seem to provide adequate counter-programmes to the challenges at hand. Here, the question of relating hope and critique comes into play. Faced with the choice between returning to more strident ideological criticism or abandoning critique altogether, critics increasingly call for (post)human forms of politics and new political imaginaries. These include affirmative and immanent reconceptualisations of critique, emphasising the importance of entanglement, multiplication, positivity and new creative possibilities (Braidotti 2018; Haraway 1997, 2008; Stengers 2011; Thiele 2015). In other words, what is asked for are the sources to create surplus: to 'transcend' given perspectives and to take up something new, a 'vertical revolution' in terms of cultivating new imaginations of the world (Cornell and Seely 2016).

## Hope's Revival

The call for new political imaginaries in the light of the end of critique corresponds to the revival of hope. In the last two decades hope has witnessed a renaissance in the context of an increased interest among social and political theorists in the utopian aspect of critical theory. Initially, this revival was informed by attempts to move beyond problems of norm justification and to think about the grounds for hoping that those principles and values considered that 'ought to govern social life will one day actually do so' (Sinnerbrink et al. 2006, 3). At its core stood the question of what it is that enables such hope (Sinnerbrink et al. 2006, 9). This revival of interest in utopia, or, more accurately, in the 'intertwined problematics of how to be utopian and how to remain hopeful' (Anderson 2006, 691), spans various fields, from sociology (Levitas 1990; 2013), to philosophy (Zournazi 2002), anthropology (Hage 2003), human geography (Anderson 2006), and ecology and environmental ethics (Kretz 2013). More recently, the issue of hope has gained particular attention in discussions about climate change and the Anthropocene (Head 2016; Lear 2006; Oram 2016; Raygorodetsky 2017). In this context, hope is seen as either a driving force behind positive change, preserving the ecosystem through 'hopeful politics' (Moellendorf 2022), or a source of 'cruel optimism' that disregards the ambivalent position of the human in the Anthropocene as signifying a departure from established notions of modernity (Berlant 2011). Concepts like *grounded hope*, rooted in reality, instead of optimism (Feldman and Kravetz 2015), or *radical hope*, as an ethics in the light of cultural devastation (Lear 2006), are meant to provide an orientation when facing the all-encompassing threat of climate change.

Some consider this revival of hope with caution. More than a decade ago, Craig Browne suspected that 'if the appeal *of* hope derives from its apparent power to restore a belief in the prospect of change, then the appeal *to* hope seems to risk undermining practical capacities for transformation' (Browne 2005, 68). Accordingly, the recent turn to hope must be understood as an ideological manifestation of a broader 'dissipation of utopian energies' (Sinnerbrink et al. 2006). The end of critique, then, coincides with hope and a reflection on why attempting to change the world is (not) futile. This resonates with concerns of advocates of hope who stress 'that we are going through a "crisis of hope"', emphasising its proximity to a more general 'crisis of narratives' (Smith 2006, 52). A second response aligns with Sarah Amsler's proposal of 'bringing hope "to crisis"', suggesting the need to develop more reflexive understandings of the interplay between the notions of 'crisis' and 'hope' (Amsler 2008). Hence, in taking a deeper look at how hope is addressed in these recent debates, the following sections analyse how doubts in the critical project do not necessarily have to be understood as a crisis of hope, but rather can be grasped as a particular way of problematising and recuperating hope. This perspective, therefore, provides starting points for repositioning critique.

Three aspects of the current revival of hope are considered to be symptomatic here. First, there is a growing discourse of *ungroundable hope* (Amsler 2008, 54f.; Rorty 1999) or *hope for the impossible* (Caputo 2003), which expresses a general scepticism towards philosophical justifications of hope. Second, there is a re-evaluation of the function of hope in terms of *not necessarily having solutions to problems,* thereby separating hope from the expectation of fulfilment (Head 2016). Lastly, there is an increased engagement with Ernst Bloch's *The Principle of Hope* (Bloch 1986), which emphasises the *affective resonances of hopeful dispositions*, seeking to cultivate an immanent 'utopianism that intervenes in the emergence' (Anderson 2006, 691). They indicate a shift from what can be called traditional forms of *utopian hope*, a goal-directed and future-oriented mode of hoping which negates the present and promulgates a better alternative, grounded upon the 'confident belief in the transformative power of collective praxis and a consciousness that human beings are self-organizing and self-determining historical agents' (Webb 2007, 78), to *radical hope*, the commitment to maximal openness to radically new possibilities under increased ontological vulnerability (Lear 2006). The following section develops insights from the broader literature and then focuses specifically on the distinct context of the Anthropocene.

## Hope's Persistence

The revival of hope would appear to work contra to depictions of the contemporary Western present as an age without hope, and in which notions of transformation and possibilities of a different future appear as naive or problematic projections.[2] Early radical re-engagements with the problem of hope attempted to develop secular practices, or an 'ethos of hope', as 'a way of embodying the conviction that the future may be different from the present' (Anderson and Fenton 2008, 77). Given the lack of telos or ultimate ground, 'one key response to the supposed removal of hope' has been to articulate hope as a transformative force, 'valued as an event that breaks with and is against that which exists' (ibid.).

Exemplary in this regard is Jonathan Lear's *Radical Hope: Ethics in the Face of Cultural Devastation*, which problematises hope in the light of the breakdown of 'a culture's sense of possibility itself' (Lear 2006, 83). Through the story of Chief Plenty Coups, the last leader of the Crow people, Lear elaborates on the ontological dimension of the possibility of the collapse of a world as the 'breakdown of the field in which occurrences occur' (Lear 2006, 34). This possibility of 'things ceasing to happen' is interpreted in terms of an 'ontological vulnerability', a vulnerability that humans share simply by being human (Lear 2006, 50). How can one live after the breakdown of one's culture, of the framework of distinctions, narratives, practices and rituals that made life meaningful? How to think with the means of this culture beyond its own end? Lear's answer is that, in this instance, hope must become radical. Radical in the sense that it is not a particular hope which determines life, but rather hope as such, the stance of a commitment to possibility, a commitment to something entirely indeterminate and beyond current understanding: 'The commitment is only to the bare possibility that, from this disaster, something good will emerge: the Crow shall somehow survive' (Lear 2006, 79). This survival does not just include 'the biological survival of the individual members of the tribe – however important that was – but the future flourishing of traditional tribal values, customs, and memories in a new context' (Lear 2006, 145). Against this background, radical hope appears as a stance of maximal openness to radically new possibilities under heightened ontological vulnerability. Hope needs to become imaginative and creative, but without knowing towards what it is directed:

> What makes hope *radical* is that it is directed towards future goodness that transcends the current ability to understand what it is. Radical hope anticipates a good for which those who have the hope as yet lack the appropriate concepts with which to understand it. (Lear 2006, 103)

Such affirmations of the necessity of hope have partly been taken up but have also shifted within debates of the Anthropocene and its radical affirmation of a break with modernity as having already occurred. Decoupled from the 'emotion of optimism', hope 'savours the life and the world we have, not the world as we wish it to be' (Head 2016, 11). There is no other world, no alternative future to build on or hope for: 'In a context of distributed agency and non-linear change it is impossible to predict where all the good or bad things will happen, or even which processes and states are likely to be more adaptive (or not)' (ibid.).

Thus, hope not only persists in the Anthropocene, it becomes pivotal in a world that no longer appears to be accessible to us (see also Chandler in Chapter 1; Waldow et al. in the Introduction). Hope persists in the urgent need to think beyond the end of the world. It persists in the affirmation of the Anthropocene as an opportunity to give up illusionary dreams of development and progress and to radically rethink human–nature relations, agency and politics within entirely new approaches to life. These approaches place hope in such possibilities as 'letting [ourselves] become more susceptible' to the world around us (Morton 2016, 129), or 'liberat[ing] us from our own mental enclosures' by thinking with forests (Kohn 2013, 22), or learning from 'the uncontrolled lives of mushrooms [. . .] when the controlled world we thought we had fails' (Tsing 2015, 2). If only we begin to embrace the deep entanglements of our current predicament, then new possibilities will emerge. As soon as we learn to acknowledge our precarious embroilment in the planetary whole, we will be able to set free our speculative imaginations – towards the world itself in its vitalising potentiality. Hope survives in the critiques of modernity: as an embrace of contingency, indeterminacy, insecurity and diversity; as a subversive capacity which is itself intrinsic to life.

This signals substantial changes in the notions of transformation and the character of anticipation. By opening up to the present and the world as it is, hope is seen to allow the cultivation of new sensibilities and forms of belonging that may contribute to reconfiguring our world beyond our current crisis. However, if there is no other world, no alternative future to build on or hope for, hope becomes an end in itself: a will or desire for life that enables governance approaches of regulation, subjectivisation or containment. Moreover, as indicated earlier, emphasis on the contingency and instability of things might foster acceptance of the status quo. Where envisioning of alternatives to existing injustices becomes problematic, hope becomes limited to reimagining what already exists. Where the emphasis is no longer upon relying on hope to transform the world, but to change our relation to the world as it is, hope labours to uphold the status quo rather than providing critique. These reconceptualisations of hope explicitly renounce the imaginaries of critical thought (and hope) as problematic human-centred modernist projections and reflect more general epistemological and ontological shifts. The focus shifts from hope to hopes, the spaces of their emergence and disappearance, to the event of hope in the here and now, to the multiplicity of hopes, relations and encounters:

> We do not know what a body that hopes does and can do. And so, what hope does varies. In one place, hope might mobilize resources, secure legitimacies, or defer justice. In another, hope opens up the present, enables bodies to keep going, or becomes part of new forms of belonging. (Anderson and Fenton 2008, 78)

These approaches attempt to free the imagination and to overcome the flaws associated with traditional utopian hope. However, they have a rather inhibiting way of constructing the relation between hope and critique, imagination, and crisis. Historically, modern forms of utopian thought motivated critique and provided orientation to social actors and political movements through envisioning alternatives to existing injustices. Today's appeals to hope are more reluctant. For some, there may still be the hope that the future may be different, but without being able to offer substantial grounds for identifying 'what it would mean that the future would be better' (Browne 2005, 70). In the Anthropocene, however, hope is restricted to an immanent re-envisioning of what is. This hope constitutes a radical break with the future which at the same time enters the present, a present in which the catastrophe has already happened, a present which appears as a present of ongoing crisis: 'an infinite repetition of a captivating state of impotentiality' (Wrangel 2018, 83), 'a radical constriction of the possibilities of the present' (Martin 2014, 63). This hope affirms the present instead of criticising it.

## An Afterword

This chapter has sought to analyse ways of articulating hope and critique together: thus, re-engaging hope as an entity able to bridge the gap between our normative claims and political practice. Claims of the end of critique are grounded precisely on this discrepancy but seek to extrapolate a more fundamental dismissal of moral or political action as not a possibility. As the chapter demonstrated, the answer to these claims is to adapt more radical, affirmative and less subject-centred approaches based upon hope that is external to us, rooted in the world of embedded and entangled relations. However, this (over)emphasis on immediate positivity can obscure the contradictions and negative aspects of our current predicament and, moreover, undermine the possibility to reflect upon the complicity of theory in sustaining them. The challenge of the Anthropocene is, then: in which direction do we orient our speculative forces? Should imagination today be confined to only the possibility of surviving by adapting to an anticipated loss? Or is a world without the privilege of hope itself a problematic promise?

The Anthropocene is a specific way of understanding our current situation, but it presents challenges during times of ongoing crises. It restricts the possibilities of the present to specific alternatives. Hope in the Anthropocene tends to be inhibiting, promoting the status quo rather than challenging it. To move beyond the abstract, wishful thinking of traditional forms of utopian hope and the empirical affirmation of existing possibilities of radical hope, hope needs to emerge from the critical analysis of power and domination of the social and material conditions within which we live. This way, hope may function as an intermediary between the extremes of complete disillusionment and hopeful affirmation, acknowledging the obstacles to moral and political engagement for a change of unjust conditions, while still not giving up on political and moral practice as meaningless. This enables the understanding of hope as a link between critique and its successful practice. It suggests that the claims of the end of critique are problematic, as they disregard their own conditions of possibility, and that we have to have hope, and thus, on a political level to engage with hope, to be able to engage in critique.

## Notes

1. In later writings, Latour expands upon his argument by introducing composition as an alternative approach to critique. This approach still involves seeking universality, but without assuming that it already exists and merely needs to be uncovered (Latour 2010, 322). Instead, he emphasises a 'tentative and precautionary progression' towards the composition of a common world that is 'much more material, much more mundane, much more immanent, much more realistic, and much more embodied' (2010, 473, 484).
2. Which might itself be a catastrophic Western-centric scenario; see Cornell and Seely (2016, 8).

## References

Adorno, T. W. 1979. Negative Dialektik. *Gesammelte Schriften 6*. Frankfurt am Main: Suhrkamp Verlag.
Adorno, T. W. 1997. Einleitung in die Musiksoziologie. *Gesammelte Schriften 14*. Frankfurt am Main: Suhrkamp Verlag.
Alt, S. 2019. Conclusion: Critique and the politics of affirmation in international relations. *Global Society* 33(1): 137–45.
Amsler, S. 2008. Bringing hope 'to crisis': Crisis thinking, ethical action and social change. Working Paper. Available at https://core.ac.uk/download/pdf/5223662.pdf (accessed 4 January 2024).
Anderson, B. 2006. Transcending without transcendence: Utopianism and an ethos of hope. *Antipode* 38(4): 691–710.
Anderson, B. and Fenton, J. 2008. Editorial introduction: Spaces of hope. *Space and Culture* 11(2): 76–80.
Austin, J. L. 2022. The public, its problems, and post-critique. *International Politics Review* (10): 92–101.
Bargués-Pedreny, P. 2019. From critique to affirmation in international relations. *Global Society* 33(1): 1–11.
Bargués-Pedreny, P., Koddenbrock K. J., Schmidt J. and Schmidt, M. (eds). 2015. *Ends of Critique*. Duisburg: Käte Hamburger Kolleg/Centre for Global Cooperation Research.
Beck, U. 2015. Emancipatory catastrophism: What does it mean to climate change and risk society? *Current Sociology* 63(1): 75–88.
Bennett, J. 2010. *Vibrant Matter: A Political Ecology of Things*. Durham, NC and London: Duke University Press.
Berlant, L. 2011. *Cruel Optimism*. Durham, NC: Duke University Press.
Bloch, E. 1986. *The Principle of Hope*, vol. 3. Oxford: Blackwell.
Braidotti, R. 2018. Ethics of joy. In R. Braidotti and M. Hlavajova (eds), *Posthuman Glossary*. London: Bloomsbury, 221–4.
Browne, C. 2005. Hope, critique, and utopia. *Critical Horizons* 6(1): 63–86.
Bunz, M., Kaiser, B. M. and Thiele, K. 2017 *Symptoms of the Planetary Condition: A Critical Vocabulary*. Lüneburg: Meson Press.
Caputo, J. D. 2003. The experience of God and the axiology of the impossible. In M. A. Wrathall (ed.), *Religion after Metaphysics*. Cambridge: Cambridge University Press, 123–45.
Chandler, D. 2015. Resilience and the 'everyday': Beyond the paradox of 'liberal peace. *Review of International Studies* 41(1): 27–48.
Chandler, D. 2016. How the world learned to stop worrying and love failure: Big data, resilience and emergent causality. *Millennium: Journal of International Studies* 44(3): 391–410.

Chandler, D. 2018. The death of hope? Affirmation in the Anthropocene. *Globalizations* 16(5): 695–706.

Colebrook, C. 2014. *Death of the Posthuman: Essays on Extinction*, vol. I. London: Open Humanities Press.

Connolly, W. E. 2011. *A World of Becoming*. Durham, NC: Duke University Press.

Connolly, W. E. 2013. *The Fragility of Things: Self-Organising Processes, Neoliberal Fantasies, and Democratic Activism*. Durham, NC: Duke University Press.

Cornell, D. and Seely, S. 2016. *The Spirit of Revolution: Beyond the Dead Ends of Man*. Cambridge: Polity Press.

Cudworth, E. and Hobden, S. 2011. *Posthuman International Relations: Complexity, Ecologism and Global Politics*. London: Zed Books.

Feldman, D. and Kravetz, L. D. 2015. *Supersurvivors: The Surprising Link Between Suffering and Success*. New York: Harper Wave.

Hage, G. 2003. *Against Paranoid Nationalism: Searching for Hope in a Shrinking Society*. Leichhardt: Merlin Press.

Haraway, D. 1997. *Modes_Witness@Second_Millenium*. London and New York: Routledge.

Haraway, D. 2008. *When Species Meet*. Minneapolis: University of Minnesota Press.

Haraway, D. J. 2015. Anthropocene, Capitalocene, Plantationocene, Chthulucene: Making kin. *Environmental Humanities* 6(1): 159–65.

Haraway, D. J. 2016. *Staying with the Trouble: Making Kin in the Chthulucene*. Durham, NC: Duke University Press.

Head, L. 2016. *Hope and Grief in the Anthropocene: Re-conceptualising Human–Nature Relations*. London: Routledge.

Hofstätter, A. 2019. Catastrophe and history: Adorno, the Anthropocene, and Beethoven's late style. *Adorno Studies* 3(1): 1–19.

Kant, I. 1998. *Critique of Pure Reason*, P. Guyer and A. W. Wood (eds). Cambridge: Cambridge University Press.

Koddenbrock, K. J. 2015. Strategies of critique in international relations: From Foucault and Latour towards Marx. *European Journal of International Relations* 21(2): 243–66.

Kohn, E. 2013. *How Forests Think: Toward an Anthropology Beyond the Human*. Oakland, CA: University of California Press.

Koselleck, R. 1982. Krise. In O. Brunner, W. Conze and R. Koselleck (eds), *Geschichtliche Grundbegriffe. Historisches Lexikon zur politisch-sozialen Sprache in Deutschland*, vol. 3. Stuttgart: Klett-Cotta.

Kretz, L. 2013. Hope in environmental philosophy. *Journal of Agricultural and Environmental Ethics* 26(5): 925–44.

Latour, B. 2004. Why has critique run out of steam? From matters of fact to matters of concern. Special issue on the Future of Critique. *Critical Inquiry* 30(2): 225–48.

Latour, B. 2005. *Reassembling the Social: An Introduction to Actor-Network-Theory*. Oxford: Oxford University Press.

Latour, B. 2010. An attempt at a 'Compositionist Manifesto.' *New Literary History* 41(3): 471–90.

Latour, B. 2014. Agency at the time of the Anthropocene. *New Literary History* 45(1): 1–18.

Latour, B. 2017. *Facing Gaia: Eight Lectures on the New Climatic Regime*. Hoboken, NJ: John Wiley & Sons.

Lear, J. 2006. *Radical Hope: Ethics in the Face of Cultural Devastation*. Cambridge, MA: Harvard University Press.

Levitas, R. 1990. *The Concept of Utopia*. London: Philip Allan Publishers.

Levitas, R. 2013. *Utopia as Method: The Imaginary Reconstitution of Society*. Basingstoke: Palgrave Macmillan.

Martin, T. 2014. Governing an unknowable future: The politics of Britain's Prevent policy. *Critical Studies on Terrorism* 7(1): 62–78.

Moellendorf, D. 2022. *Mobilizing Hope: Climate Change and Global Poverty in the 21st Century*. Oxford: Oxford University Press.

Morgan, A. 2019. Reconciliation with nature: Adorno on reason, nature and critique. *Adorno Studies* 3(1): 20–32.

Morton, T. 2013. *Hyperobjects: Philosophy and Ecology after the End of the World*. Minneapolis: University of Minnesota Press.

Morton, T. 2016. *Dark Ecology: For a Logic of Future Co-Existence*. New York: Columbia University Press.

Oram, R. 2016. *Three Cities: Seeking Hope in the Anthropocene*. Wellington, NZ: Bridget Williams Books.

Pugh, J. 2020. The affirmational turn to ontology in the Anthropocene: A critique. In M. Stephens and Y. Martínez-San Miguel (eds), *Contemporary Archipelagic Thinking: Toward New Comparative Methodologies and Disciplinary Formations*. Lanham, MD: Rowman & Littlefield, 65–83.

Raygorodetsky, G. 2017. *The Archipelago of Hope: Wisdom and Resilience from the Edge of Climate Change*. New York: Pegasus.

Rorty, R. 1999. *Philosophy and Social Hope*. Harmondsworth: Penguin.

Schmidt, M. and Koddenbrock, K. 2019. Against understanding: The techniques of shock and awe in Jesuit theology, neoliberal thought and Timothy Morton's philosophy of hyperobjects. *Global Society* 33(1): 66–82.

Sinnerbrink, R., Deranty, J.-P. and Smith, N. H. 2006. Critique, hope, power: Challenges of contemporary critical theory. In R. Sinnerbrink,

J.-P. Deranty, N. Smith and P. Schmiegen (eds), *Critique Today*. Leiden and Boston, MA: Brill Academic Publishers, 1–21.

Shore, M. 2017. A pre-history of post-truth, East and West. *Eurozine*. Available at https://www.eurozine.com/a-pre-history-of-post-truth-east-and-west/ (accessed 4 January 2024).

Smith, N. H. 2006. Hope and critical theory. In R. Sinnerbrink, J-P. Deranty, N. Smith and P. Schmiegen (eds) *Critique Today*. Leiden/Boston: Brill Academic Publishers, 45–63.

Stengers, I. 2011. *Thinking with Whitehead: A Free and Wild Creation of Concepts*. Cambridge, MA: Harvard University Press.

Stengers, I. 2015. *In Catastrophic Times: Resisting the Coming Barbarism*. London: Open Humanities Press.

Thiele, K. 2015. Ende der Kritik? Kritisches Denken heute. In A. Allerkamp, P. Valdivia Orozco and S. Witt (eds), *Gegen/Stand der Kritik*. Zurich and Berlin: diaphanes, 139–62.

Toscano, A. 2012. Seeing it whole: Staging totality in social theory and art. *Live Methods* 60(S1): 64–83.

Tsing, A. L. 2015. *The Mushroom at the End of the World: On the Possibility of Life in Capitalist Ruins*. Princeton, NJ: Princeton University Press.

Webb, D. 2007. Modes of hoping. *History of the Human Sciences* 20(3): 65–83.

Wesche, T. 2012. Moral und Glück. Hoffnung bei Kant und Adorno. *Deutsche Zeitschrift für Philosophie* 60(1): 49–71.

Wrangel, C. T. 2018. The Use of Hope: Biopolitics of Security during the Obama Presidency. PhD dissertation, University of Gothenburg.

Zournazi, M. (ed.). 2002. *Hope: New Philosophies for Change*. Annandale, NSW: Pluto Press.

# 12

# Hope in a World That Will Never End? The Problem of Fanatical Hope in Critical Dystopias

*Aristidis V. Agoglossakis Foley*

**Introduction**

Dystopia, though a relatively recent term (Sargent 2006), dates back to archetypal images of hell and the bleakest possible imaginable futures (Claeys 2015). Dystopian literature in its wider sense, as a mode of expression for these speculative, depressing futures, finds its modern origins towards the end of the nineteenth century and with the works of Edward Bulwer-Lytton and H. G. Wells. Since then, especially after the 1920s and Yevgeny Zamyatin's influential *We*, there was an unmistakable boom in the amount of dystopian fiction produced, not only in literature, but also in film, television and art. Most dystopian authors provide a rather thinly veiled criticism of their contemporary social and political practices and norms, by projecting them onto the speculative canvas of potential futures. Authors address themes such as totalitarianism, technological innovation and ecological catastrophe, amongst others, to provide a potential map of the future that can be used to criticise and comment on their current world. That is why examining dystopias, both past and present, can provide an interesting analysis of contemporary phenomena. The narratives and themes in dystopian fiction can provide a starting point against which socio-political phenomena can be compared, giving us the ability to deem such practices 'dystopian', as mass media frequently does (Gilbert 2016; Poole 2019). Furthermore, the study of dystopian thought can provide insight similar to the study of the Anthropocene, the wider theme of this book, due to their shared rejection of modernity and the centrality of themes such as rapid technological growth, ecological catastrophe and overpopulation. However, dystopia provides a bleaker perspective, as it can be thought of as the natural result of the Anthropocene, imposing a sense of finality. Within this dystopian milieu, this chapter will address the importance, position and dubious necessity of a relatively secondary dystopian theme: hope.

The theme of hope appears in two ways. First, there are dystopias in which hope is a prevalent theme. It is contrasted with the grim reality presented by the authors, giving readers a 'way out' and negating the totalising oppression of those dystopias. These, as we shall see further on, maintain an important role in the overall genre. Second, there is a trend in dystopian studies, championed by Tom Moylan and Rafaella Baccolini, to search for hope in dystopias whether it exists within the narratives *or* without. Critical dystopias embody a zealous and incessant search for hope, understanding the concept as key to the study of dystopias (Moylan and Baccolini 2003). Out of the many conceptions of hope that exist (Wahyuna and Fitriana 2020), Ernst Bloch's theory has dominated both the utopian and dystopian schools of thought. Understanding hope as 'a direct act of a cognitive kind' (Bloch 1996, 12), *The Principle of Hope* portrays it as a necessary and inherent human emotion, as it is 'the most human of all mental feelings' (Bloch 1996, 75), which drives us to search for a better future. This understanding of hope is quite utopian and therefore maintains a key role in utopian studies (Levitas 1990), though less so in the dystopian tradition.

Thus, this chapter explores how such a conception of hope, intrinsically linked to Bloch's utopian work, can be attached to the study of dystopian thought. To conduct this analysis, three distinct conceptions of hope will be examined. Bloch's domination of the dystopian tradition, through critical dystopias, will be explored first. I argue that this dogmatic imposition of hope, applied with a zealous fervour, can appear to be almost fanatical. Following fanatical hope, the next two conceptions emerge from the field of classics and have been chosen because they neither understand, nor position, hope in the same way with regard to the human psyche as Bloch. The second conception of hope that this chapter uses, negative hope, stems from Thucydides' understanding of hope as a fool's game, the very last option left to a doomed individual. Finally, drawing from Aristotle's far more optimistic notion of hope, which is linked to courage and bravery, the chapter examines dystopias in which positive hope plays a key role in the narrative.

The argument of this chapter has two objectives. First, it seeks to examine how critical dystopias' search for hope ceases to portray its original Blochian, utopian, characteristics, instead devolving into a fanatical search for hope. This may lead the act itself of studying hope within dystopia to resemble the negative hope that Thucydides warned of: meaningless hope in the face of impending doom. The argument builds on Davison-Vecchione's point that critical dystopias neuter the critical ability of traditional dystopias that do not maintain a utopian impulse, as 'dystopia's critical import becomes, so to speak, parasitic on utopia' (Davison-Vecchione 2021). The search for a utopian, Blochian, hope dominates the power of the so-called critical dystopias, leading scholars of critical dystopias to embody the act of fanatical hope. Second, if hope must be analysed, the overreliance on the Blochian conception of hope provides a rather limited analysis. Introducing other conceptions, such as the positive and negative ones derived from Aristotle and Thucydides respectively, allows for a more nuanced study of the theme. Overall, the chapter argues that the search for Blochian hope in the dystopian genre may actually prove detrimental to the critical power that defines it. If hope can be seen to provide a way out, inducing readers to a blissful lethargy, then any criticism embedded in the dystopian works is dampened. This impacts the wider conception of the dystopia's psyche, as such a powerful concept may, gradually, lose its ability to prompt contemplation and criticism of contemporary socio-political practices.

Before proceeding, it is important to briefly address the dystopias included in this chapter. Instead of a detailed textual analysis, I track each conception of hope throughout the last two centuries. More importantly, I have included many Indigenous or non-Western dystopias, which are very often overlooked and overshadowed by Western 'classics'. Their inclusion allows for the exploration of positive and negative hope based on Indigenous realities. Furthermore, the issue of definition must also be addressed. Though numerous understandings of dystopia and anti-utopia have emerged in the past fifty years, the dystopias of this chapter follow this specific definition: dystopias are forms of critical expression, primarily literary, that propose a distorted, negative, image of the near future. This definition combines the critical intent and centrality of contemporary political and social norms found in Raymond Williams's definition (Williams 1979, 63), with Lyman Tower Sargent's understanding that dystopia is an 'extrapolation from the present that involves a warning' (Sargent 1994, 8). Combining the two, dystopias provide criticism and warning derived from the political and social context of each author.

## Fanatical Hope

The concept of hope which Bloch proposes in *The Principle of Hope* differs immensely from the two 'classic' conceptions that will follow. For Aristotle and Thucydides, hope maintains its fundamental characteristics as a passive human trait. For Bloch, however, hope represents an inherently human need, an action. Rather than just being an emotion, the need to hope for an improved situation or even for a better world is, for Bloch, intrinsic to the human psyche (Bloch principle). Naturally, this very positive understanding of hope is well suited to the study of utopian thought. Contrasting with negative perceptions of utopia, such as Karl Popper's (Popper 2002), Bloch holds it as the gold standard to which one can compare and 'judge existing practice', leading us away from totalitarianism and closer to a more positive future (Sargent 1994). This idea of utopia, which holds spiritual and ritualistic characteristics, is influenced by Bloch's own religious and Marxist background (Volf 1986), as a promise and search for the good place, the eutopia, personified in images of heaven and stateless communism.

It is Bloch's conception of hope that is being applied to the study of dystopia. As Patricia Vieira points out, there is a trend in the field of dystopian studies to 'classify the various types of dystopias based upon their relation to Ernst Bloch's famous "principle of hope"' (Vieira 2020, 353). Originally, Sargent defined critical dystopias as dystopias that include 'at least one eutopian enclave', fostering hope for the destruction of the imagined dystopian reality to be 'replaced with a eutopia' (Sargent 2001, 22). Echoing Sargent's previous idea of a hopeful warning (Sargent 1994, 26), critical dystopias clearly include elements that allow for the hope of a utopia. Following in Sargent's steps, Tom Moylan and Raffaella Baccolini both focus a considerable amount of their work to further establish critical dystopias.

However, their understanding of critical dystopias has developed Sargent's original conception. Mirroring critical utopias – that is, utopias that are aware 'of the limitations of the utopian tradition' (Moylan 2014, 10–11) – critical dystopias are aware of their dystopian limitations, including some forms of utopian hope. They contrast traditional dystopias, which 'maintain utopian hope *outside* their pages, if at all' (Moylan and Baccolini 2003, 7) to critical dystopias, in which 'the ambiguous, open endings of these novels maintain the utopian impulse *within* the work' (Moylan and Baccolini 2003, 7). Thus, an odd dichotomy is created in which hope maintains a key role, as it can exist both within and outside dystopian texts, critical or otherwise. This exploration of hope both in the pages of dystopias and their milieu imitates the Blochian necessity to hope. Further exploring the connection of hope to the dystopian genre and critical dystopias, Thaler argues that hope and despair are far more connected than we may think (Thaler 2019). By bridging the two, one presupposes that where despair dominates, hope must not be far behind. Through the establishment of critical dystopias, hope emerges as a key theme and as part of the dystopian thought.

It is at this point, at least with regard to the field of dystopia, that the conception of fanatical hope can be used to provide a critical analysis. Fanatical hope is not just an understanding of hope in itself, but rather, it simultaneously mirrors, mutates and satirises Bloch's conception, thus resembling an aggressive human act. It should be understood as the practice, the incessant need, to find hope anywhere and everywhere, to be constantly hopeful, even when hope may be neither helpful not present. Drawing from Thucydides' understanding of negative hope, where hope resembles a trait that only appears before damnation, fanatical hope is used to criticise these attempts to find hope, as they can be considered a last-ditch attempt for salvation in the face of oblivion.

I argue that this relentless, fanatical search for hope is precisely the trap into which scholars who work on critical dystopias fall into. As stated in Moylan and Baccolini's (2003) key work on critical dystopias, *Dark Horizons: Science Fiction and the Dystopian Imaginations*, hope is found in the ambiguity of the open endings of critical dystopias, allowing the reader to imagine how the story may progress, and what the future of the inhabitants may be. This leads to two interlinked problems. First, hope is not part of the narrative, but rather subject to the imagination of each reader. Second, by imposing this purely subjective hope onto the overall narrative that relies on an ambiguous ending, other themes that these dystopias may be critical of are overshadowed.

Such analyses are prevalent in *Scraps of the Untainted Sky* (2000), where Moylan provides a lengthy examination of critical dystopias. Out of the wide breadth of literature he draws from, two texts prove to be relevant for this argument: *The Machine Stops* by E. M. Forster and Yevgeny Zamyatin's *We*. Admittedly, in both cases, an argument for hope is possible. Forster's dystopia ends with the inexplicable destruction of the Machine, and *We* ends on an ambiguous note where, though the main character, D-503, has essentially been lobotomised, a revolution is underway. However, in both cases, whether hope exists or not is subjective, and neither signifies a utopian impulse. The entirety of the underground human population perishes along with the Machine, the destruction of which is not necessarily an act of hopeful resistance. By contrast, though resistance is prevalent at the end of Zamyatin's dystopia, the fate of D-503 is despairing and the outcome of the revolution is unknown. Whichever argument one may choose to make, the real problem is that by focusing on hope, other prominent themes are overlooked. Rising and dominating technology, lack of identity and individuality, oppression, the relationship between humanity and nature, and inter-human relations are all powerful themes in both (Forster 2011; Zamyatin 1993). More importantly, they are products of criticism based on concrete socio-political phenomena in each author's context, as opposed to the ambiguously open-ended subjective possibility of hope. Therefore, though the argument for hope can be made, it should not be done in a way that eclipses the strong socio-political criticism inherent in these dystopias.

Though the original conception of critical dystopias as narratives that include hope can provide an interesting insight into slightly more 'hopeful' dystopias, the application of a Blochian understanding and constant need to search for hope may not be the most beneficial for the study of dystopian thought. Primarily through the works of Moylan and Baccolini, among others, the persistent search for hope, even if absent, dampens and blunts the critical power of each dystopia. If hope is to be found, dystopias' inherent criticism is diminished, providing the reader with a hopeful lifeline, a salvation that the author may not have desired. In a bizarre twist, by relentlessly searching for hope we may end up as the hopeful fools that Thucydides warned against – hoping foolishly, blindly, in the face of total destruction, thus drawing our efforts and attention away from taking action or providing any meaningful solution.

**Negative Hope**

Originating from *The History of the Peloponnesian War*, Thucydides' conception of hope can be predominantly found in the Melian Dialogue. Differing from the rest of his historical account, which is presented as a series of speeches made by individuals, the Melian Dialogue is written as a debate or dialogue between the Athenians and the Melians as singular entities. Considered a 'masterpiece within a masterpiece' (Wassermann 1947, 18), it provides insight into the socio-political and philosophical differences that distinguish Athens and Melos, and it is often used as a clear example of political realism (Crane 1998). In their pre-siege negotiation, the Athenians urge the Melians to surrender for their own self-preservation. Thucydides defines hope as 'that comforter in danger' (Thucydides 1972, 404), also translated as 'danger's comforter' (Schlosser 2012); thus the Athenians warn the Melians not to place their chance for a future in hope, but rather to act and submit. Reflecting Thucydides' own ideals, the Athenians embody a very negative image and understanding of hope. They deem hope to be 'among the most dangerous political decisions', as it signifies the 'only possible response to despair' (Schlosser 2012, 169). They go so far as to accuse the Melians of relying on superstition and abstractions, resting their future on 'prophecies and oracles and such things which by encouraging hope lead men to ruin' (Thucydides 1972, 404). This portrayal of a negative understanding of hope extends beyond the Melian Dialogue. The tragic characters of Pericles and Nicias both show hopeful aspirations in the face of ruin, but both fail miserably. Even the Athenians, in a turn of classic irony, find their hopes of victory destroyed following the Sicilian Expedition (Schlosser 2012, 180). Thus, for Thucydides hope signifies a fool's errand, a course taken when all is lost. It represents the last laughable stab in the dark before total destruction.

This purely negative understanding and conception of hope is one that can be found in many dystopias from the past century, born out of different contexts and critically addressing separate issues. Two of the three 'classic' dystopian novels of the first half of the twentieth century, Aldous Huxley's *Brave New World* and George Orwell's *Nineteen Eighty-Four*, present narratives in which hope appears as a very bleak, foolish concept. In Huxley's dystopia, the reader is confronted with a society divided according to a caste system, in which oppressive technological innovation, rampant consumerism and unlimited distractions all work to suppress and tame the individual. Originating from Huxley's fear that 'people will come to love their oppression' and 'adore the technologies that undo their capacities to think' (Postman 2006:19), combined with a disgust for consumerism (Combs 2008), the themes of governmental subjugation and technology dominate this dystopia (Popova 2012).

However, hope, in a very Thucydidean sense, is also prevalent in the narrative. Following Bernard Marx, an Alpha-Plus disillusioned with the seemingly free system, the reader is introduced to John the Savage. Born outside the rigid caste system, John provides the perfect example of Thucydides' conception of hope. Through John, Huxley cultivates the idea that hope may be possible in his dystopia. However, by the end of the novel, John succumbs to state conditioning and commits suicide, highlighting the complete domination of the regime (Huxley 2007). It becomes clear that any hope that may have existed was not only superficial or false, but also a final failed aspiration to change the reality of *Brave New World*. In other words, to believe that John may be able to keep his individuality and live outside the controlling clutches of the regime represents hope as Thucydides understands it. Never real, it is a final cry for help before utter despair.

Equally, Orwell's dystopia cultivates false hope, though some may disagree (Atwood 2005; Claeys 2017; Winter 1983). In the first half of *Nineteen Eighty-Four*, Orwell alludes to the possibility of true hope, only to destroy it in the second half. The attempts of Winston Smith, the main character, to go against the Party and Big Brother seem foolish, indeed acting as a 'comforter in danger'. Winston's and his lover Julia's complete submission to the state and Party dogma emphasises the complete and utter hopelessness of their position (Orwell 2013). Addressing similar themes to Huxley, Orwell's novel stems from criticism and fear of both the Nazi regime and the Soviet Union. These dystopias, published before the 1950s, highlight the inability to escape the total control of the two regimes. The hope that is presented in each novel dissolves into hopelessness and conformity: the perennial boot stamps out all hope in the end (Orwell 2013, 307).

John Christopher's 1956 *The Death of Grass* continues this Thucydidean conception, examining humanity's disastrous blunders, hoping irrationally to alter the course of its destruction. Addressing themes of ecological catastrophe, the atomic bomb and racial othering (Watt 2015), the novel depicts Western attempts to combat a Chinese crop-killing virus in the 'focal point of the Anthropocene' (Tongsukkaeng 2020, 214). Choosing to bomb major cities in a failed attempt to stop the virus, Christopher highlights the inability to avoid the impending doom. Representing this through ecological catastrophe and human-made destruction, the actions that humanity takes, such as strict quarantine or the use of hydrogen bombs, signify attempts for false hope as a last resort (Christopher 1970). Like the Melians, humanity's choice to place hope in irrational actions, in Christopher's dystopia, leads them to their doom.

Yoko Ogawa's *The Memory Police*, published in 1994, provides a completely different narrative to the three preceding dystopias, but it maintains a similar understanding of hope. Set on an island, this semi-Kafkaesque novel focuses on the power of memory and explores its impact on both collective society and individuality. Addressing themes reminiscent of the Japanese government's attempt to conceal and ignore the fallout of the Second World War, the inhabitants of Ogawa's dystopia live under the control of the Memory Police. As the novel progresses, different aspects of nature, human society and the individual psyche are forcefully forgotten, culminating in the disappearance of individual body parts and finally the erasure of the self (Ogawa 2020). Negative, foolish hope is found in the character R, who is unaffected by the tactics of the Memory Police. After the main character fades into inexistence, R emerges from their hiding place, and the novel ends. Though this decision to potentially confront the totalitarian regime could be seen as an act of courage, it is an act of desperation. With no other way out remaining, R is faced with complete defeat. Emerging from the cave is an act of hope born in despair, a final attempt at salvation before destruction.

Moving to the twenty-first century, Margaret Atwood's short story 'Time Capsule Found on the Dead Planet' provides the first recent instance of Thucydidean hope. Written in anticipation of the 2009 Copenhagen Climate Change Conference (Atwood 2009), it imagines the complete evolution of the human species, culminating in our unavoidably self-imposed extinction. Criticising the role that capitalism and human greed play in the impending ecological doom, Atwood addressed the hopeless attempts of humanity to find hope in such a condition. After the desertification of the globe, efforts to construct a positive, almost hopeful, narrative are made: 'a stone in the sand in the setting sun could be beautiful' (Atwood 2009). This false hope in the face of extinction represents the human folly that hoping against hope entails (see Zebrowski in Chapter 2).

A very different depiction of false hope is present in Dave Eggers's 2013 *The Circle*. Exploring themes of technocracy and corporatocracy (Tuters 2015),

Eggers sets his contemporary dystopia in the near future. The Circle, the dominating technological company (and thinly veiled as a nod to Google), enforces an Orwellian motto onto its consumers: 'Secrets are Lies. Sharing is Caring. Privacy is Theft' (Eggers 2013). Following Mae, a new employee of the Circle, the reader explores the dystopia that unfolds. Constant surveillance, lack of individualisation, corporate domination and greed are all strong themes of the novel. Similarly, to Orwell and Huxley, Eggers creates the illusion of hope. However, false hope emanates from a secondary, revolutionary, character, Ty Gospodinov, who is betrayed by Mae (Eggers 2013). By constructing hope in such a way, Eggers emphasises the truly hopeless nature of his dystopia. Gospodinov's attempt to combat the dominating corporate regime is a last attempt at hope before complete subjugation. Mae's actions demolish any hope, stressing the hopeless aspect of *The Circle* through the reader's forced identification with her.

*Tender Is the Flesh* by Agustina Bazterrica, published in 2017, provides a rather gruesome conclusion to this section. Addressing climate issues and meat consumption, cannibalism becomes legal in this dystopia following a virus that has spoiled all nonhuman meat. Marcos, a married human-meat farmer, slowly develops a relationship with a female member of his human cattle, Jasmine. As the novel progresses, and the relationship between Marcos and Jasmine becomes more human, hope is kindled in the reader. Humanity seems salvageable, as the main character increasingly humanises the 'other'. However, after Jasmine gives birth, Marcos takes her to be slaughtered, stating that 'she had the human look of a domesticated animal' (Bazterrica 2020). Any hope of absolution is shattered, as Marcos and his wife used Jasmine to provide them with a child because the couple was unable to conceive. Similar to the other dystopias of this section, a false hope is created for the reader, only to lead to certain doom. Much like for the Melians before them, hope in these dystopias is based on falsehoods and wishful thinking, signifying acts of utter despair.

## Positive Hope

Strongly contrasting hope understood as a fool's errand, Aristotle's conception provides a far more positive outlook. In *The Nicomachean Ethics* and *Rhetoric*, two conceptions of hope emerge. The former links hope with courage and confidence, whereas the latter provides an understanding of hope as a product of good fortune. This section uses the first understanding, the hopeful triptych, as Aristotle disregarded hope as good fortune, as a mere 'confidence in the positive outcome of the circumstances' (Gravlee 2000, 465). Hope is central to bravery, according to Aristotle, providing fertile ground for courage, and underlying 'the deliberation and self-confidence necessary both to improve one's circumstances and to cultivate the excellence of character' (Gravlee 2000, 477). This centrality of positive hope nurtures and promotes bravery and confidence, both of which are 'a mark of a hopeful disposition' (Aristotle 1996). This conception of hope, which stands in stark contrast to the Thucydidean negative understanding, can also be found in the pages of dystopian fiction. This section points to the ways in which hope is constructed throughout the narratives, and, more importantly, survives until the very end. This courageous hope in dystopia provides the possibility for true, meaningful, change. Mirroring the previous section, a limited exploration of hope explores how this positive understanding can be found in the dystopias of the past century.

Starting with the first of the three aforementioned 'classic dystopias', Yevgeny Zamyatin's *We* provides the archetypal map for modern dystopian fiction. Influencing Orwell (Claeys 2010), Zamyatin's dystopia was written in 1921 but not published in English until 1924, being banned in Russia until 1989 (Zamyatin 1970). Zamyatin used the narrative of *We* to criticise the relatively new bureaucracy and totalitarian tendencies of the Bolshevik regime. His 'most jesting and serious work' (Zamyatin 1970, 4), the dystopia explores a world in which individuality is lost and the state dictates every aspect of social life. All the buildings are made of glass, sex is organised, and each day is strictly regulated. However, within this narrative Aristotelian hope is everywhere, presenting a different understanding from the hope that Moylan imposes onto the ambiguous ending. Following the life of D-503, an engineer of the *Integral*, a spaceship designed to export the One State's totalitarianism, Zamyatin slowly sows the seeds of dissent. Through a narrative similar to *Nineteen Eighty-Four*, with the introduction of a female character, I-330, a rebel group, the Mephi, and a world beyond the 'Green Wall', hope starts to blossom. At the end of the dystopia, the Green Wall, which separates the technological regime from nature, is crumbling. Furthermore, I-330 does not succumb to her torture, though D-503 has undergone brainwashing, losing his free will. Despite the demise of the main character, the courageous resistance of I-330 and the destruction of the wall create hope. Drawing from hope, the Mephi cultivate courage and confidence, highlighting the possibility of resistance (Zamyatin 1993). Thus, in this novel that addresses both the alarming power of technology and the totality of the state (Lynskey 2021), Aristotelian hope can be found in the end.

In a similar manner to *We*, courageous hope can be found in *Fahrenheit 451* by Ray Bradbury. Published in 1953, this famous dystopia explores themes of technology, oppression and book burning. A 'book of warning' (Gaiman 2013), this dystopia criticises new modes of entertainment, addressing the rising popularity of television and McCarthyist censorship. Montague, a firefighter who burns books, slowly begins to realise the vacant superfluity of the world around him. By reading prohibited books, he becomes enlightened to his position. After the destruction of his city, Montague, with a group of intellectuals, returns to help rebuild, using the knowledge they preserved from illegal books (Bradbury 2008). Though the dystopian setting is unnerving, brought into being by the population (rather than the actions of a totalitarian regime), hope is clear in Bradbury's dystopia. The retention of knowledge and the possibility of rebuilding both show the reader that the dystopian condition is finite. Through hope, courageous acts of defiance are possible and can allow for change.

Moving to 1979–80, two dystopias, from Somalia and Kenya, address the themes of courageous hope and resistance. The first instalment of his 'Variations on the Theme of an African Dictatorship' series, Nuruddin Farah's *Sweet and Sour Milk* was written during his self-imposed exile from Somalia. Presenting a Kafkaesque dystopian setting with crime fiction traits, the novel explores a murder. Loyaan, whose twin is the victim, sets out to discover the murderer, gradually uncovering secrets of his brother's revolutionary actions (Farah 2006). The reader is gradually introduced to the reality of a neocolonial dystopia, in which the state dominates and the colonial influence is prevalent. An allegory of late 1970s Somalia, the totalitarian president displays similarities with Joseph Stalin and the long-reigning Somali president, Major General Barre (Brooker 1995, 62). Despite Farah's thinly veiled criticism of the neocolonial state, through the actions of his dead brother Loyaan begins to develop hope and courage. He becomes involved, slowly taking his brother's place in the resistance. Farah uses Loyaan's gradual acceptance of his role to prove that hope and opposition against the dominant regime are possible.

In a similar manner, Ngugi wa Thiong'o's *Devil on the Cross* addresses the theme of neocolonialism and totalitarianism while providing Aristotelian hope at the end. Originally written in Gikuye and published only two years after President Jomo Kenyatta's death, the novel includes deeply anti-colonial and Marxist ideals, while simultaneously being very critical of Kenyatta's regime. Written during his imprisonment for his criticisms of the postcolonial government (Thiong'o 2018), wa Thiong'o's critique of the government is symbolised by the use of the term 'haraambe' – used by Kenyatta to signify community but becoming a synonym for neocolonialism and corruption in the novel. Set in Kenya, Wariinga, left pregnant by the 'Rich Old Man', is introduced to the 'Devil's Feast', a bourgeois celebration of colonial and working-class exploitation where Western businesses and local delegates mix. Utilising Christian imagery, Thiong'o emphasises the evil character of the oppressors. However, despite the total control that these individuals and businesses have over the population and the state, hope is a prominent theme in his dystopia. At the end of the novel, two years after the 'Devil's Feast', Wariinga resists the neocolonial condition she finds herself in by killing the 'Rich Old Man' and other feast participants (Thiong'o 1987). Similar to Loyaan's, this violent act shows courage and hope against the dominant regime, and an ability to oppose it. Both courageous acts are performed to improve the condition of each character.

Two recent dystopias also express the courageous hope that Aristotle proposes in *The Nicomachean Ethics*. Published in 2013, Basma Abdel Aziz's *The Queue* acts as a vessel for her criticism of the Egyptian government, the Muslim Brotherhood and the extreme bureaucracy of that period. Set in an unnamed Middle Eastern country, the dystopia is inspired by a queue that Aziz witnessed, lasting for two hours in front of a closed government building, and the religious and military rule dominant at the time (Underwood 2016). The story unfolds in front of the 'Gate', a bureaucratic building that processes paperwork requesting permission to perform *any* action. Focusing on a doctor and his patient, readers explore the dystopia over 140 days, as the doctor waits to get the necessary approval that permits him to perform his healing duties. Finally, after 140 days of waiting, the doctor goes against the rules and attempts to save his patient – though it is too late (Abdel Aziz 2019). Although this dystopia may initially seem devoid of hope, the actions of the doctor show courage and hope in the face of totalitarian oppression. Even though the patient is lost, and the doctor waited too long, the act of defiance shows that hope is present, allowing him to perform the courageous act of insubordination, personifying the brave individual Aristotle wrote of.

Finally, hope in 2017's *Autonomous*, by Annalee Newitz, exists in the actions of the main character, Jack. An eco-dystopia, Newitz uses the novel to criticise current failed attempts to avoid the climate crisis, along with the domination of Big Pharma and corporatocracy. The Robin Hood-type character of Jack provides subscriptions illegally for the poor. Simultaneously, he must avoid agents sent to silence him, following his discovery that a pharmaceutical drug causes individuals to become addicted to their work (Newitz 2018). Through Jack, the dystopia explores both capitalist issues surrounding the cost of medicine, cheap labour and the ability of an individual to resist. Like most of the dystopias in this section, courageous hope is found in the actions of the main character. Unlike the previous section, the fate of these individuals is not doomed. Rather, their attempts to resist the regime are valid, legitimising the hope they create. In these dystopias, each author wished to include some courageous hope, giving individuals the confidence to resist totalitarian oppression.

## Conclusion

Hope in the study of dystopian thought maintains a very important position. Both its inclusion and exclusion from a narrative highlights different, crucial aspects of the intended dystopian criticism. Non-Western, postcolonial dystopias, such as *Devil on the Cross* and *Sweet and Sour Milk*, though very critical and clearly against neocolonial regimes, provide hope to emphasise that resistance is possible. On the other side of the coin, dystopias that predominantly deal with the theme of technology, like *Brave New World* and *The Circle*, provide speculative narratives in which hopefulness is mocked and hope is deceptive. However, through the analysis of critical dystopias, hope is being searched for in the genre of dystopia in a fanatical manner. Finding hope in ambiguous open endings, scholars of critical dystopias fall into the trap of searching for vague hope, ignoring the concrete criticisms included in 'critical' dystopias.

Following the analysis of fanatical hope, I have attempted to provide alternatives to the dominant Blochian study of hope in dystopia by exploring two different conceptions. Whether positive or negative, hope is largely restrained to the pages of each dystopia, limited to its textual form. Here, by adopting multiple perceptions of hope, we can track how it appears over the past two centuries, linking the conception of hope within these texts to the historical contexts from which it emerged. Additionally, it is important to examine hope not within a vacuum, but posit it in relation to the other themes that each dystopia addresses – whether it be technology, ecology, propaganda, amongst others. In this way, we can study the theme of hope in dystopia in a far less dogmatic, yet simultaneously multifaceted, manner, allowing for more pluralistic conceptions of dystopia, despair and hope.

## References

Abdel Aziz, B. 2019. *The Queue*. London: Melville House Publishing.

Aristotle. 1996. *The Nicomachean Ethics*. H. Rackham (trans.). Stansted: Wordsworth Editions.

Atwood, M. 2005. *Curious Pursuits: Occasional Writing 1970–2005*. London: Virago.

Atwood, M. 2009. 'Time Capsule Found on the Dead Planet' by Margaret Atwood. *The Guardian*, 25 September. Available at https://www.theguardian.com/books/2009/sep/26/margaret-atwood-mini-science-fiction (accessed 5 January 2024).

Bazterrica, A. M. 2020. *Tender Is the Flesh*. New York: Scribner.

Bloch, E. 1996. *The Principle of Hope*, vol 1. Cambridge, MA: MIT Press.

Bradbury, R. 2008. *Fahrenheit 451*. London: HarperCollins.

Brooker, M. K. 1995. African literature in the world system: Dystopian fiction, collective experience, and the postcolonial condition. *Research in African Literatures* 26(4): 58–75.

Christopher, J. 1970. *The Death of Grass*. London: Penguin.

Claeys, G. 2010. The origins of dystopia: Wells, Huxley and Orwell. In G. Claeys (ed.), *The Cambridge Companion to Utopian Literature*. Cambridge: Cambridge University Press, 107–31.

Claeys, G. 2015. Three variants on the concept of dystopia. In F. Vieira (ed.), *Dystopia(n) Matters*. Newcastle upon Tyne: Cambridge Scholars Publishing.

Claeys, G. 2017. *Dystopia: A Natural History: A Study of Modern Despotism, Its Antecedents, and Its Literary Diffractions*. Oxford: Oxford University Press.

Combs, R. 2008. The eternal now of *Brave New World*: Huxley, Joseph Campbell, and *The Parennial Philosophy*. In D. G. Izzo and K. Kirkpatrick (eds), *Huxley's Brave New World: Essays*. New York: McFarland.

Crane, G. 1998. *Thucydides and the Ancient Simplicity: The Limits of Political Realism*. Berkeley, CA: University of California Press.

Davison-Vecchione, D. 2021. Dystopia and social theory. *Ideology Theory Practice*, 18 October. Available at http://www.ideology-theory-practice.org/1/post/2021/10/dystopia-and-social-theory.html (accessed 5 January 2024).

Eggers, D. 2013. *The Circle*. New York: Vintage.

Farah, N. 2006. *Sweet and Sour Milk*. Minneapolis, MN: Graywolf Press.

Forster, E. M. 2011. *The Machine Stops*. London: Penguin Classics.

Gaiman, N. 2013. Introduction. In R. Bradbury, *Fahrenheit 451*. London: Simon & Schuster.

Gilbert, S. 2016. Donald Trump's dystopias. *The Atlantic*, 8 November. Available at https://www.theatlantic.com/entertainment/archive/2016/11/donald-trumps-dystopias/506975/ (accessed 5 January 2024).

Gravlee, G. S. 2000. Aristotle on hope. *Journal of the History of Philosophy* 38(4): 461–77.

Huxley, A. 2007. *Brave New World*. New York: Vintage.

Levitas, R. 1990. Educated hope: Ernst Bloch on abstract and concrete utopia. *Utopian Studies* 1(2): 13–26.

Lynskey, D. 2021. The 100-year-old fiction that predicted today. *BBC Culture*, 3 September. Available at https://www.bbc.com/culture/article/20210902-the-100-year-old-fiction-that-predicted-today (accessed 5 January 2024).

Moylan, T. 2000. *Scraps of the Untainted Sky: Science Fiction, Utopia, Dystopia*. Boulder, CO: Westview Press.

Moylan, T. 2014. *Demand the Impossible: Science Fiction and the Utopian Imagination*. R. Baccolini (ed.). New York: Peter Lang.

Moylan, T. and Baccolini, R. 2003. Introduction: Dystopia and histories. In T. Moylan and R. Baccolini (eds), *Dark Horizons: Science Fiction and the Dystopian Imagination*. Abingdon: Routledge, 1–12.

Newitz, A. 2018. *Autonomous*. London: Orbit.
Ogawa, Y. 2020. *The Memory Police*. New York: Vintage.
Orwell, G. 2013. *Nineteen Eighty-Four*. London: Penguin.
Poole, S. 2019. Wake up, humanity! A hi-tech dystopian future is not inevitable. *The Guardian*, 18 February. Available at https://www.theguardian.com/commentisfree/2019/feb/18/technological-progress-superjumbo-airbus-dystopia-future (accessed 5 January 2024).
Popova, M. 2012. Happy birthday, Aldous Huxley: A rare, prophetic 1958 interview. *The Atlantic*, 26 July. Available at https://www.theatlantic.com/entertainment/archive/2012/07/happy-birthday-aldous-huxley-a-rare-prophetic-1958-interview/260369/ (accessed 5 January 2024).
Popper, K. 2002. *The Open Society and Its Enemies*, vol. 1: *The Spell of Plato*. Abingdon: Routledge Classics.
Postman, N. 2006. *Amusing Ourselves to Death: Public Discourse in the Age of Show Business*. London: Penguin.
Sargent, L. T. 1994. The three faces of utopianism revisited. *Utopian Studies* 5(1): 1–37.
Sargent, L. T. 2001. US utopias in the 1980s and 1990s: Self-fashioning in a world of multiple identities. In P. Spinozzi (ed.), *Utopianism/ Literary Utopias and National Cultural Identities: A Comparative Perspective*. Bologna: University of Bologna.
Sargent, L. T. 2006. In defense of utopia. *Diogenes* 53(1): 11–17. https://doi.org/10.1177/0392192106062432.
Schlosser, J. A. 2012. 'Hope, danger's comforter': Thucydides, hope, politics. *The Journal of Politics* 75(1): 169–82. https://doi.org/10.1017/S0022381612000941.
Thaler, M. 2019. Bleak dreams, not nightmares: Critical dystopias and the necessity of melancholic hope. *Constellations* 26(4): 607–22. https://doi.org/10.1111/1467-8675.12401.
Thiong'o, N. wa. 1987. *Devil on the Cross*. Portsmouth, NH: Heinemann.
Thiong'o, N. wa. 2018. *Wrestling with the Devil: A Prison Memoir*. New York: Vintage.
Thucydides. 1972. *History of the Peloponnesian War*. R. Warner (trans.). London: Penguin.
Tongsukkaeng. 2020. When grasses and grains were ravaged: Reading the viral pandemic and natural disaster in science fiction, *The Death of Grass*. *Journal of Human Sciences* 21(2): 214–34.
Tuters, M. 2015. Through glass darkly: On Google's gnostic governance. In D. Berry and M. Dieter (eds), *Postdigital Aesthetics: Art, Computation and Design*. New York: Palgrave Macmillan, 245–58.

Underwood, A. 2016. Waiting for Egypt's next revolution. *Los Angeles Review of Books*, 2016. Available at https://lareviewofbooks.org/article/waiting-egypts-next-revolution/ (accessed 16 January 2024).

Vieira, P. 2020. Utopia and dystopia in the age of the Anthropocene. *Esboços: Histórias Em Contextos Globais* 27(46): 350–65.

Volf, M. 1986. An interview with Jürgen Moltmann. In G. McLeod Bryan (ed.), *Communities of Faith and Radical Discipleship*. Macon, GA: Mercer University.

Wahyuna, A. H. and Fitriana, S. 2020. The concept of hope in the Western and Eastern perspective. *International Journal of Islamic Educational Psychology* 1(1): 22–36.

Wassermann, F. M. 1947. The Melian Dialogue. *Transactions and Proceedings of the American Philological Association* 78: 18–36.

Watt, K. 2015. Fear of the East: Othering in the *Man in the High Castle* and *Death of Grass*. Katalina Watt [blog]. URL no longer available.

Williams, R. 1979. Utopia and science fiction. In P. Parrinder (ed.), *Science Fiction: A Critical Guide*. Abingdon: Routledge.

Winter, F. 1983. Was Orwell a secret optimist? The narrative function of the 'Appendix' to *Nineteen Eighty-Four*. In B. J. Suykerbuyk (ed.), *Essays from Oceania and Eurasia: George Orwell and 1984*. Antwerp: Progressef, 79–89.

Zamyatin, Y. 1970. *A Soviet Heretic: Essays by Yevgeny Zamyatin*. M. Ginsburg (trans.). Chicago, IL: University of Chicago Press.

Zamyatin, Y. 1993. *We*. London: Penguin Classics.

## 13

# Hope Makes Strange: Affect, Hope and Strangeness

*Srishti Malaviya*

**Introduction**

In this chapter I interrogate the sense of strangeness that the affective becoming of hope illuminates. I argue that in taking-form hope illustrates the strange invisible entanglements that compose phenomena and make it multiple. My objective is to highlight the affective ongoingness of phenomena that interminably haunts the present. I emphasise here the concept of strangeness to highlight how imaginations of a finished, knowable world conceal the imperceptible presences and forces that continually compose the world into fluctuating and indeterminate shapes and constellations which cannot be predicted in advance. This framing is at odds with literature on the Anthropocene, which places an unambiguous emphasis on the pronouncement of 'ends'– the end of international relations (Burke et al. 2016), of the world (Harrington 2016), of history (Hamilton 2018), of time (Fagan 2019) and of hope itself (Chandler 2019). These proliferating conceptions of ends demonstrate the Anthropocene's tendency to dismiss continuity in favour of finitude while manufacturing a world in crisis. In this crisis-ridden world which needs to start anew, hope is resuscitated as a necessary value that can be consciously applied to overcome ends. Counter to the Anthropocene imaginary, the hope–strangeness relation I want to map does not have a beginning or an after, nor does it correspond to a finite conception of the world. Following Ben Anderson's framing of hope as an affect, I explore the strangeness that both inhabits the concept of hope and is unravelled as a quality of becoming in the emergence of hope. This strangeness challenges the Anthropocene's most pressing concern – of living together in the world 'after its end' – as it brings to the fore the ongoingness of the world and the always already collective becoming that sustains this ongoingness.

The chapter proceeds in the following manner: in the next section I look at what it means for hope to be read through affect theory, followed in the subsequent section by an exploration of the strangeness that illuminates its becoming. I begin by briefly discussing the conceptual contours of affect theory and then proceed to discuss Anderson's conceptualisation of hope as affect and the implications of this framing. I argue that, most importantly, Anderson's framing enables us to encounter the immanent strangeness that unsettles secure understandings of experience and phenomena, and thus unmoors hope from pre-given subjects. I then proceed to further explore this unmoored hope in its latent strangeness by identifying two immaterial aspects of it – withinness and witness – following Erin Manning's discussion of the minor gesture and Karen Barad's discussion of quantum entanglements, respectively. I conclude by briefly reflecting on how this strangeness enables us to not only recognise but also conceptually reinvigorate the indeterminacy that is at the heart of ongoing experience.

**Affect, Body and Movement**

Outlining affect first demands a rethinking of the body – a body that is different from the modernist individual body which packs within its determinate boundaries an individualised subject acting in and upon a given world, and also from the body of representational discourses that attach subjectivity to markers of identity in a socially constructed space and time. The body here is, first, what Erin Manning (2006, xiii) calls a 'processual body'. The processual body is an ever moving and transforming accretion of forces, matters and potentialities, which is always in the process of formation; a body that is continually composing and decomposing in interactions and intra-actions with other bodies in space and time. Thus, no single body is ever a given stable entity or exists in insolation; even inanimate bodies weather and decompose – intra-act with the environment – over differing timescales.

Thus, the focus here shifts from asking 'what a body is' to 'what a body can do, what they can become, and how listening to these relational possibilities might reinvigorate a politics of invention' (Blackman 2009, 135). This concept of the processual body posits body as ever in motion and thus never concretely given: 'when a body is in motion, it does not coincide with itself. It coincides with its own transition, its own variation' (Massumi 2002a, 4). A body is thus composed of its trajectories of transition as it moves in space-time. Every 'body' is alive with the potential to move and come into relation. This dynamic notion of the body is thoroughly relational. Bodies are moving constellations of relations, emerging out of relational composition with other bodies and the environments they move in and shape. Manning reminds us that this relational body

> is not a 'new' body (though it always networks in new and diverging ways). It is a body that has always emerged through and alongside other bodies, be they political bodies, gendered bodies, raced bodies. What is 'new' about the body [. . .] is not its shape or form but the relational matrices it makes possible. (2006, xiii)

That is, with the foregrounding of the relational body it becomes possible to interrogate relations and compositions that have hitherto remained outside theoretical grasp but have always existed beyond the foreclosed regimes of the signified, over-coded body. In the many ways that bodies move, and through this movement intra-act and interact with space and other bodies, relations are always shrinking, expanding, welling and overspilling. Indeed, this conception of the processual body alters what has become so fundamentally, and unalterably, identified as the 'human'. As Nigel Thrift (2008, 2) asserts, 'there is no stable "human" experience because the human sensorium is constantly being re-invented as the body continually adds parts into itself; therefore, how and what is experienced as experience is itself variable'.

At the heart of this variation in the experience of bodies in spacetime are the intensive forces of affect. In its most basic definition, affect is the capacity of a body to affect and be affected (Massumi 2002a; 2002b). Affect is the uncapturable, ungraspable gamut of 'forces' that compel a body to move, become, alter and respond. These are

> visceral forces beneath, alongside, or generally other than conscious knowing, vital forces insisting beyond emotion – that can serve to drive us toward movement, toward thought and extension, that can likewise suspend us (as if in neutral) across a barely registering accretion of force-relations, or that can even leave us overwhelmed by the world's apparent intractability. (Seigworth and Gregg 2010, 1)

Affect thus transpires as an intensive force, passing just beneath the material surface of bodies, and hence remaining imperceptible. It extends forth from encounter towards response. It always already involves more than one body, since it transpires *in-between* stimulus and action, encounter and response, bodies and spacetime, but it transpires without registering itself to conscious recognition – 'something that happens too quickly to have happened', inhabiting, and moving from within the complex nexus of possibilities into the becoming of the event (Massumi 2002a, 30). Affect imperceptibly operates in the domain of the virtual potentialities of bodies, producing 'a space of intensity that overflows a represented world' of subjects and objects (Anderson 2009, 79).

As such, affect is indeterminate, it is 'the unassimilable' intensity that registers change (Massumi 2002a, 27). It can neither be traced on surfaces, nor attributed to individual subjects, nor predicted or calculated in advance. Affect does not belong to any one body, in the same way that emotions or feelings can be attached or attributed to bodies. Affect moves non-consciously; it is the invisible movement of forces that enable action, occurrence, displacement, moving experience from one actualisation to the next. In this manner affect is 'unqualified [. . .] not ownable or recognizable and it is thus resistant to critique' (Massumi 2002a, 28) and to technologies of capture. The affective puts new relations of becoming into motion, compelling, and compelled by, 'as yet unknown powers' of the body to affect and be affected (Hardt 2007, x). Thus, the affective ushers in a conceptualisation of what Deleuze and Guattari (2005) have called an asignifying and asubjectifying politics. Affect does not speak of given subjects and their traceable displacements but *moves* relations, making it impossible for bodies to intransigently inhabit determinate positions of subjectivity when movement is infinitely recomposing relations.

It is arguably in this sense that Massumi (2002b, 212) thinks of affect as the word he would use for 'hope' – it registers the value of the relational, the indeterminate, the imperceptible and the potential. It does not confine experience to structurally defined parameters of possibility but engenders the 'nextness' of movement and the newness immanent to phenomena (Manning 2009). Here hope does not reside *in* a particular outcome or in the anticipation of change or even in an identified subject – it is a moving force that reorients analysis to 'the emergent and the prospective' (Anderson 2017, 595).

## The Unmooring of Hope

Taking off from this treatment of hope through affect as the point of departure, Ben Anderson interrogates how hope takes place, examining the immanent provisions of its mode of becoming through a specific focus on Bloch's conception of hope as anticipating the 'not-yet become' (Bloch 1986, cited in Anderson 2006, 733). For Anderson (2006), hope is a relation that emerges from particular encounters within a present infused with 'diminishments', and hints at a yet-to-form that is beyond or 'outside' the horizon of present experience. The two concepts of 'diminishments' and 'outside' situate Anderson's (2006) 'geographies of hope' in the domain of the affective, wherein the co-presence of paradoxical potentials propels a sudden disruption in the disposition of bodies and inducts into the ongoingness of experience the possibility of difference. By 'diminishments' Anderson means the constraints that operate on the spacetime of the present and that encumber a body and take away from the experience of the present. These diminishments set up the possibility for hope in the first place, which Anderson (2006, 743, emphasis added) calls 'a point of danger or hazard' wherein 'the present is *haunted* by the fact that something good that exceeds it has yet to take place and the "conditions that make it possible to hope are strictly the same as those that make it possible to despair" (Marcel, 1965, page 101)'. Hope, thus, brings about a paroxysmal paradox as it intensifies the experience of present diminishments but yields from this intensification a sudden sense of what is yet to come, and what is yet to be acted towards (see also Waldow in Chapter 11).

This indeterminate yet-to-come arrives as an intimation from the beyond or 'outside' of the horizon of present diminishments, drawing forth 'a point of contingency within a present space-time' (Anderson 2006, 744). This contingency expresses itself as a possibility of difference in the ways in which the present spacetime is felt and experienced. The arrival of this disruption further allows bodies to apprehend the multiple presences and heterogeneities that make up for the composition of spacetime, which exceed present constraints and are outside of 'what has already become in "everyday life"' (Anderson 2006, 746). Thus, hope surges as an intangible movement between what is and what remains yet to come. There can be no exact and direct reciprocity between the hopeful disposition and the actualised outcomes of the movement of bodies, affects and spacetimes. The actual cannot completely correspond to the infinitude of affective dispositions and so hope as an affect cannot be measured by the outcomes it may or may not translate into. What is more valuable in this framing are the tendencies and capacities that hope illuminates, in a moment confounded by constraints and despair, leading to a qualitative transformation in the felt experience of the present.

Hope, thus, brings about in bodies the capacity to affect and be affected, propelling trans-formation in a 'proprioceptive and visceral sense' (Anderson 2006, 743). In this sense, hope is affective. While it is not tethered to particular narratives of change or a given content of projections for the future, it emerges from in between the restraints of the present and the yet-unknown difference of unfinished, ongoing becoming; it causes a rift in the spacetime of experience and 'enables bodies to go on' (Anderson 2006, 744). Once again, the movement that hope engenders cannot be plotted across explicit coordinates of positioned experience. Instead, it is felt in the qualitative, infinitesimal shifts in relations, both of the body with itself and of the body with its immediate situational context, its spacetime, and other bodies in the spacetime. It is also noteworthy here that Anderson's framing of 'other bodies' exceeds the human and even the tangible. His case studies specifically focus on the relational nexus that takes form in conversation with music. He explores how the relations between his respondents, their afflicted contexts, and the sudden irruption of music into each of the respondents' immediate experience, as well as in their conversations with Anderson, bring about a hopeful shift in their disposition towards their present diminishments.

Two significant implications emerge out of hope's affective framing. First, hope is unmoored from a specifically anthropocentric vocabulary of particular desires, aspirations and conundrums. That is, hope is detached from its overcoded iterations as a temporally indexed sentiment which mediates with the ways in which an individual or collective subject narrativises and represents the past, present and future as discrete categories of experience. Affective hope cannot be possessed by a body or referentially attributed to conscious ownership by a body. It is not something that bodies have or do not have; it is that which can cause bodies to change and become or experience their environment differently. Ceaselessly perpetuating itself outside the boundaries of linguistic representation, and alongside the multiple becoming of bodies and spacetimes, hope is felt in and as sensation. It is sensed in the promising surge of difference within a present moment. It is sensed as a call from something beyond the topological figuring of what is. It is sensed in the movement between diminishments and difference. It is sensed as a 'relation of suspension' between an open-ended future and a capacity to 'dwell more intensely in points of divergence within encounters that diminish' (Anderson 2006, 747).

Second, this haunting suspension opens up a spectral margin for the sensing body between its present paradoxes and diminishments, and the unfinished progression of the process of living. This differential margin is 'spectral' because it looms impalpably, not available to tangible approximation. It is the strangeness of this spectral margin that I propose to engage with further here. Anderson's 'geographies of hope' are haunted by strange presences and forces that circulate with and within bodies. Engagement with this strangeness can have important consequences for rethinking the ways in which 'living together' is conceptualised and realised, displacing the human standpoint as the singular position from which this togetherness is mostly defined. The unmooring of hope from human subjectivity is crucial in order to truly pluralise and transversalise the concept, to free it from the universalising constraints of a necessarily human narrative and make it a matter of ongoing experience itself. In this unmooring, what turns inside out is precisely the subterranean, unfathomable strangeness that always already haunts the ways in which spacetime is composed and experienced. This strange haunting draws to the fore the sense of that which simultaneously, infinitely moves from within and moves together-with.

## Strange Becomings

Thinking through the affective registers an electrifying passage of strangeness on two interrelated levels. At first comes the 'shock to thought', following the sudden plunge of language, representation, subjectivity and determinate certainty into shape-shifting, plural incipience that is fundamentally at odds with the static, grounded and singularly defined articulations of modern life (Massumi 2006). The affective brings about a radically startling departure from the known forms and designations that life, body and the world remain arrested within under enduring regimes of signification and identification. At this level, the shock is conceptual in that it anticipates a radical rethinking of dominant paradigms of knowing and being. On a second level, wherein the conceptual shock is sensuously felt, are the encounters with strangeness in the flickering, fleeting, ephemeral and mobile figurations of experience that become available to sensing bodies in the wake of affective movements. The affective hints at entanglements which persist just beyond the horizon of the perceptible, and which register a qualitative change in the experience of the present, but nonetheless pass in movement before they can be tangibly captured or grasped.

Hope as an affect intensifies this strangeness – it heightens the experience of liminality between what is and the unknown potentialities that could transform into what is yet to happen. By extension, hope illuminates the liminal figuring of the body as being beyond the possession of one-self. This indeterminacy, suddenly available to the senses in the wake of hope, disturbs settled meanings of what is here and there, inside and outside, now and then. Nicholas Royle's (2003, 1) articulation of the uncanny in *The Uncanny* offers an un/settling account of the strangeness of the 'peculiar commingling of the familiar and the unfamiliar'. For Royle (2003, 1) the uncanny ushers in 'a crisis of the proper [. . .] of the natural, touching upon everything that one might have thought was "part of nature": one's own nature, human nature, nature of reality and of the world'. This is a crisis of the senses precipitated by the impossibility of closure, the unstoppability of becoming, the indefinability of phenomena. The uncanny is the sudden, almost infinitesimal and hence unconfirmable, spilling over of the underlying strangeness of becoming into perceived phenomena, a split fraction of a moment of becoming privy to the impossible ongoingness of life in its thorough enmeshment with bodies, animate and inanimate, material and sensual, and with spacetime in its unrelenting movement of becoming. This moment of the uncanny perpetuates 'a sense of ourselves as double, split, at odds with ourselves' (Royle 2003, 6). 'I' am suddenly not my-self; 'I' am inseparably enmeshed in phenomena both seen and unseen and known and unknown. And yet this enmeshment is not of an order that can be physically highlighted, demonstrated or proven. The strangeness of this becoming more than one-self is the problematisation of origin and ends, the negation of the fixed knowledge of beginnings and endings, in the continual becoming and un-becoming of the world. The uncanny reveals the thorough strangeness of the constant churning of invisible forces of intra-action, continually gathering in and dispersing out, and which sustain the ongoing becoming of worlds. As the dizzying motion of bodies in simultaneous becoming and worlds in unyielding motion is registered in its passing, the uncanny strangeness arrives as a 'flickering sense (but not a conviction)' (Royle 2003, 1).

The uncanny of hope begins precisely in what Anderson recognises as its haunting of the present, the disruption it effects in felt experience of the present without a corresponding material change. Hope brings an inkling of the world's ongoing processuality which unsettles the perceived fixity of present conditions. This intimation not only highlights the fluid composition of experience and body, but also illuminates the virtual spectrality of potential. That is, it brings about an awareness of the presences that coexist within us and with us, unseen and ungrasped but real, populating and shaping the world. These presences are spectral not only because they remain out of reach of tangibility and capture, but also because they are always looming, always already present, entirely ungovernable. But most of all the strangest possibility of hope as affect is the recognition of the ways these spectral presences are not 'outside' of 'us'; indeed they make the 'outside' unrecognisable in the becoming of 'us' by making inside/outside a distinction impossible to sustain. That is, the unceasing ingress of the not-yet-become into the ways in which present diminishments are encountered makes the hopeful disposition animate a resistant strangeness that defies categorical representations of space, time and self. It instead operationalises 'a mode of originality that either cannot be assimilated, or that so assimilates us that we cease to see it as strange' (Bloom 1982, cited in Royle 2003, 15). The uncanny strangeness of hope either remains inexhaustibly ungovernable, or it folds us into its ungovernability so that we can no longer sense ourselves outside of it. In either case, it persists. I identify here two simultaneous facets of this resistant strangeness: on the one hand, hope attunes experience to its constant variability that irrupts from within the ongoing movement of experience itself. On the other hand, this variability takes form by gathering within its force-fields that which exceeds it, yielding 'a single event that is not one' (Barad 2010, 244). I now turn to mapping out these strange realms of hope in the two relations of 'within' and 'together' through Erin Manning's concept of the minor gesture and Karen Barad's concept of cutting together/apart, respectively.

## *Beyond Within*

Erin Manning's discussion of the minor gesture, drawn from Deleuze and Guattari's concept of the minor laid out in *A Thousand Plateaus*, expands on the internal variability of experience. The minor gesture is 'the gestural force that opens experience to its potential variation [. . .] from within experience itself, activating a shift in tone, a difference in quality' (Manning 2016, 1). The minor is the affectual 'withinness of movement', the force-field of relations that enable materialisation of the actual (Manning 2009). The minor subsists within the major, which is a name for the seemingly permanent and given structural forces which organise experience into knowable categories and discourses, 'according to predetermined definitions of value' (Manning 2016, 1). The major organises experience into discrete units of being and produces knowledge about this organisation. The major is thus the plane of identification, naming, sovereignty; the plane on which things and phenomena become 'real' or concretised into tangible, structural knowledge. The major is the site where change is registered, measured, valued based on a set of norms already identified. Change is captured within the narrow frame of displacement from the structural, organisational norm and ascribed a fixed value. The major is the site where space is identified as territory, and time is quantified in indifferent and measurable units. Manning contends that the major is identified as the stage where change happens not because it is the source of transformation, 'but because it is easier to identify major shifts than to catalogue the nuanced rhythms of the minor' (2016, 1).

Nonetheless, it is the gestural force of the minor that continually moves experience and spacetime into recomposing figurations that may or may not concretise into measurable movements. The minor gesture protrudes out of the affective interplay of potentialities within phenomena, often enabling shifts in the concrete realm. The minor itself remains unmeasurable, unpredictable, indeterminate. The minor sustains itself, produces its own value, passes into nothingness, and becomes again, anew, something else. The minor sustains the becoming of continuity, punctuating the process of becoming (Manning 2016, 3). This punctuation is precisely what makes the minor a gesture – from within the folds of experience it affects the progress of process, ceaselessly recomposing the limits of experience by always expressing difference, and always expressing differently, 'inventing new modes of life', making strange the major definitions of space, time, experience (Manning 2016, 8).

For instance, Manning identifies neurotypicality as a major organiser of experience, tending to the production of identifiable and governable forms of perception. Taking the example of autistic perception, Manning (2016, 14) discusses how this form of perception destabilises the major not so much by the quantitative identification and naming of neurodivergence but by virtue of 'the opening, in perception, to the unclassifiable, the ungovernable'. Autistic perception does not instantaneously extrapolate identifiable forms of experience from the infinitude of sensations that populate the world. Instead, it is attuned to this infinitude, to the potential that congeals into recognisable figurations, directly perceiving the complexity of worlds in motion, what Manning (2014) has elsewhere referred to as 'wondering the world directly'. This is unlike the majoritarian norm of neurotypical perception wherein a body readily superimposes given forms on the unfolding of worlds. The autistic perception can sense the ongoing variation within experience, making its force-field of relational becoming 'other-worldly', attuned to the ceaseless invention of worlds in motion (Manning 2016, 191). This other-worldliness is not what makes autistic perception different – the otherworldliness of experience is contained within its infinite potential for internal variability. It is the capacity to register this continual other-worldly invention of worlds that colours the strangeness of autistic perception.

Anderson's framing of the hopeful disposition, as the capacity to suddenly perceive the ongoingness of experience, and its capacity for internal variation identified in the yet-to-come, make this strangeness an integral part of affective hope. Hope is a minor gesture in the sense that it is invariably at odds with the congealed constraints that limit the body-becoming in the passing present, and yet from this oddness it produces a capacity to go on, act towards change. It arises unpredictably and passes into bodies, creating ripples in the fabric of the spacetime of experience. It may or may not produce change on the plane of the major, yet it occasions a shift both in the particular ecology the body is immersed in, and in the body's capacity to act. It simultaneously illuminates the difference that is always already looming, just beyond the horizon of perceptibility, and occasions a reaching-towards this beyond. Hope does not predict in advance what the beyond consists of, nor does it make available a specific content that populates the beyond. Instead, the terrain of hope is the capacity to reach-forward, to keep moving towards that which is yet-to-come. It brings about perception and awareness of this beyond, which is both not-yet-here but also arises from that which is already here, already contained in experience. Hope folds the experience of spacetime in ways that distort notions of linearity and givenness, opening it to its own simultaneous immersion in the past, present and future. Perpetually reaching-towards the beyond, while negotiating the constraints that resist that beyond, hope is a minor tendency that plays to the rhythms of what Manning recognises as 'spectral politics' – a politics that never reaches its 'goal' because it is always tending to the continuing of movement further, of reaching-towards further (Manning 2006, 138–40). Spectrality colours hope: 'its happening is always a kind of un-happening. Its 'un' unsettles time and space, order and sense' (Royle 2003, 2).

*Never Alone*

A body is never alone. It does not become alone; it does not act alone. A body's becoming is always haunted by the co-presence of affective intensities that surround it invisibly, act upon it immaterially and flow from it endlessly. Bodies are not so much entangled with other clearly defined bodies and phenomena; they become in entanglement with other bodies and phenomena in spacetime. Thus, a body is always haunted by the ways in which other bodies and phenomena become, while its own becoming haunts the materialisation of other bodies and phenomena. This is Karen Barad's concept of intra-action, which replaces the notion of interaction where prefigured bodies independently act with each other to create measurable outcomes. Intra-action precedes bodies, and 'the co-constitution of determinately bounded and propertied entities results from specific intra-actions' (Barad 2010, 253).

The entanglements that perform the work of intra-action are always multiple, simultaneous, paradoxical and, in Barad's (2010, 265) terms, 'hauntological', whereby the 'enfolded traces of othering' are present within bodies. Barad (2010, 244) draws on quantum entanglements, specifically the ways in which 'electrons experience the world', to demonstrate the uncanniness of how matter is 'never one with itself'. This specifically has to do with a certain kind of plurality that coheres, multiplies, interferes, diverges and disappears in no particular order, in the taking-form of matter, and that is composed of particles that make a leap *'without having been anywhere in between'* (Barad 2010, 246, emphasis in original). This dis/joint movement makes phenomena internally multiple and disparate, the multiplicity stitched together across contingent lines of disjuncture, at once 'differentiating and entangling' (Barad 2010, 265). Barad calls this 'cutting together/apart' – the movement of quantum entanglement, making the singular composed of multiple. That is, 'one' is always many, always more than one. Massumi has referred to it as 'ever-varying manyness of all that comes as one' (Massumi, in Manning 2013, ix).

Thus, entanglement is a relational ecology with countless ghostly propensities and potentialities always moving just beneath the surface of the 'one', composing the one in response to each other and to the environment. The capacity to respond thus does not begin with an individual, self-contained agent of action, nor does it have an origin prior to the becoming of a body, a singularity, a whole. At once, it happens in the process of becoming and makes becoming possible. This is the 'dis/continuous' movement of entanglement, where Barad (2010) echoes Derrida who echoes Hamlet in saying 'time is out of joint', in response to the dis/array of presences that populate spacetime.

In everyday spacetime, this quantum entanglement can be fleetingly sensed in the affective movement that takes shape in the passage of hope. Overlayed with the 'flickering sense' of the uncanny, the passage of hope reveals the indeterminate patching together of disparate phenomena and entities, as it becomes in the registering of an excess that exceeds the present singularity, even as it arises from it. Hope 'folds into the being together of corporealities to disrupt, momentarily, the circulation of despair through the ingression of [. . .] an-other' (Anderson 2006, 743). The 'trace of othering', the composite fractions of ongoing intra-action, are essential to the becoming of hope. Hope takes-place both with the other and in the recognition of an-other, dissolving the categorical separation of self and other, and composing an ever-differentiating sameness (Barad 2010). Compositional co-presences are sensed as hope courses through a body, leaving a trace of the 'strangeness of framing and borders' (Royle 2003, 2).

Further, the hopeful body senses the hauntological 'inheritance' of its 'entangled relationalities', hauntological because the inheritance is not simply a matter of being passed down in a linear passage of ownership, but of the body emerging from and with the haunting persistence of these entangled relationalities (Barad 2010, 264). The movement between moving states of becoming emerges from the ongoing response of bodies to the relational fields they are entangled in, and hope capacitates a body to open up to 'indeterminacy in moving towards what is to come' (Barad 2010, 264). Hope concretises into the feeling of an ability to respond, not in particular outcomes of response. Hope thus attunes a body to the 'ongoing responsibility to the entangled other' (Barad 2010, 265). It enlivens the work of intra-action, where bodies are not acting upon one another but become in the threading together of diverse affects and presences in a compositional conversation, such that the hauntological relational becomes a field of responsive figurations. Hope then induces a body to intensely experience the uncanny presences of the ghostly, the immaterial, the not present, nor absent, the strange, gather them into ongoing response and 'inhabit uncertainty, together' (Massumi 2002b, 218). It estranges a body from clearly demarcated categories of past, present and future, here and there, now and then, and plunges it into 'the continual reopening and unsettling of what might yet be, of what was, and what comes to be' (Barad 2010, 264). Cutting together-apart, hope pierces a body with the manyness that composes it, re-splicing it together in the capacity to act.

## Conclusion

The unmooring of hope from human subjectivity, and the consequent unveiling of the uncanny nature of affective forces that compose and decompose becoming, serves to highlight the value of indeterminacy and the impossibility of discrete experience. The foundational indeterminacy within existence and experience poses a significant challenge to the idea of a purposive world, by foregrounding an idea of an ongoing world that is not predetermined towards human contentment. It does so while also explicitly making clear that this indeterminacy is not a novel 'crisis' emerging out of an anthropocentric world remade in the image of the human, as the notion of the Anthropocene suggests, and in which hope is restricted to being a tool to achieving desired outcomes for living together in the world after its proclaimed end. Radical indeterminacy, and its strange compartments, has always already informed the ways in which experience takes form. Reading hope as an affect allows for this indeterminacy to be recognised. The hauntological richness of hope is precisely in this capacity to register the multiplicity of phenomena that it flushes bodies with. The capacity to sense strangeness situates the spectral margin of indeterminacy at the heart of experience, making encounters with the extraordinary a necessary component of experience.

## References

Anderson, B. 2006. Becoming and being hopeful: Towards a theory of affect. *Environment and Planning D: Society and Space* 24(5): 733–53.
Anderson, B. 2009. Affective atmospheres. *Emotion, Space and Society* 2: 77–81.
Anderson, B. 2017. Hope and micropolitics. *Environment and Planning D: Society and Space* 35(4): 593–5.
Barad, K. 2010. Quantum entanglements and hauntological relations of inheritance: Dis/continuities, spacetime enfoldings, and justice-to-come. *Derrida Today* 3(2): 240–68.
Blackman, L. 2009. Starting over: Politics, hope, movement. *Theory, Culture & Society:* 26(1): 134–43.
Bloch, E. 1986. *The Principle of Hope*. Oxford: Blackwell.
Bloom, H. 1982. Freud and the sublime: A catastrophic theory of creativity. In *Agon: Towards a Theory of Revisionism*. Oxford: Oxford University Press.

Burke, A. et al. 2016. Planet politics: A manifesto from the end of IR. *Millennium: Journal of International Studies* 44(3): 499–523.

Chandler, D. 2019. The death of hope? Affirmation in the Anthropocene. *Globalizations* 16(5): 695–706.

Deleuze, G. and Guattari, F. 2005. *A Thousand Plateaus: Capitalism and Schizophrenia*. Minneapolis and London: University of Minnesota Press.

Fagan, M. 2019. On the Dangers of an Anthropocene Epoch: Geological Time, Political Time and Post-human Politics. *Political Geography* 70: 55–63.

Hamilton, S. 2018. Foucault's end of history: The temporality of governmentality and its end in the Anthropocene. *Millennium: Journal of International Studies* 46(3): 371–95.

Hardt, M. 2007 Foreword: What affects are good for. In P. T. Clough and J. Halley (eds), *The Affective Turn: Theorizing the Social*. Durham, NC: Duke University Press.

Harrington, C. 2016. The ends of the world: International relations and the Anthropocene. *Millennium* 44(3): 478–98.

Manning, E. 2006. *Politics of Touch: Sense, Movement, Sovereignty*. Minneapolis: University of Minnesota Press.

Manning, E. 2009. *Relationscapes: Movement, Art, Philosophy*. Cambridge and London: MIT Press.

Manning, E. 2013. *Always More than One: Individuation's Dance*. Durham, NC: Duke University Press.

Manning, E. 2014. Wondering the world directly – or how movement outruns the subject. *Body & Society* 20(3–4): 162–88.

Manning, E. 2016. *The Minor Gesture*. Durham, NC: Duke University Press.

Marcel, G. 1965. *Being and Having*. Glasgow: Collins.

Massumi, B. 2002a. *Parables for the Virtual: Movement, Affect, Sensation*. Durham, NC: Duke University Press.

Massumi, B. 2002b. Navigating movements: A conversation with Brian Massumi. In M. Zournazi (ed.), *Hope: New Philosophies for Change*. Annandale, NSW: Pluto Press, 210–43.

Massumi, B. 2006. *A Shock to Thought: Expression after Deleuze and Guattari*. London: Routledge.

Massumi, B. 2013. Prelude. In E. Manning, *Always More than One: Individuation's Dance*. Durham, NC: Duke University Press, ix–xxiii.

Royle, N. 2003. *The Uncanny*. Manchester: Manchester University Press.

Seigworth, G. J. and Gregg, M. 2010. *The Affect Theory Reader*. Durham NC: Duke University Press.

Thrift, N. 2008. *Non-Representational Theory: Space, Politics, Affect*. London and New York: Routledge.

# 14

# Reimagining Hopeful Anthropocene Futures: From Entanglements to Radical Openness

*Ignasi Torrent*

**Introduction**

Over the last two decades, coinciding with the popularisation of conversations on the Anthropocene, namely the geological era of the anthropogenic footprint on ecosystems, various disciplines in the natural and social sciences, including earth system sciences, biology, philosophy and anthropology, amongst others, have increasingly pushed to the fore analysis of the relational ontogenesis of being. Barely questioned, this focus on the entangled feature of world elements potentially offers a mode of living with the current ecological and socio-political planetary crisis. In short, a dominant framing has taken hold of hope in the Anthropocene, one which fetishises a presumed relational condition of being, that forms the object of interrogation in this chapter. This plethora of accounts suggest that the refiguration of durable relations between living species, human and nonhuman, and their material environments might lead to emancipatory futures (see Haraway 2016). Instead, the analysis here depicts this full-blown relationality as problematically deterministic and hints at radical openness as a mode of opening up new, more creative possibilities for hope in the Anthropocene.

Heavily influenced by this broad trend, the social sciences reflect this reinvigorated interest in this relational character of being to project hopeful modes of becoming in the Anthropocene. The disciplinary proneness towards the exploration of entanglements, relations, assemblages, interconnections, entwinements, knots, frictions, collisions, negotiations and so on as constitutive of reality has been celebratorily called, by Milja Kurki (2020, 112), a 'relational revolution'. Indeed, a significant body of literature, from a wide range of sensitivities, substantiates this thematic of a relational ontology of entanglement (see Castro Pereira and Saramago 2020; Tucker 2018; Wendt 2015; Zanotti 2019). For these entangled formulations, being is always generated in relation.

This chapter seeks to reimagine a mode of hope in the Anthropocene that surpasses a universalising entangled ontology. It does not intend one more account of what is defined here as entanglement fetishism, namely the celebratory projection of the world as an all-encompassing relational wholeness (see Torrent 2021). Instead, seeking to push an important set of criticisms that have exposed the limits of this all-out relationality (see Colebrook 2019), the present chapter highlights a problematic ontological assumption often overlooked by both entanglement fetishists and their critics: that entanglements are infallibly generative, they deterministically precipitate further beings and events. In doing so, the text invites contemporary scholarship to explore non-generative encounters, of what Gilles Deleuze (1993, 59), through Leibniz, calls 'incompossibility'.

Thus, the first analytical line of the chapter hints that the preconditions for being ontologically relational are not already given, addressing the question about the possibility of being without being in relation. Drawing from a rather unorthodox line of research (see Glissant 1997), this chapter unearths non-relational, or beyond-the-relational, instances, whose prime form of being only becomes relational through the logics of subjugation, as illustrated below. For this mode of being, the second argument of the text suggests, non-engagement, refusal and withdrawal become forms of political resistance and survival, thus distorting the conventional association between political subjectivity and emancipation (see Ferguson 2021). That is, the chapter proposes that hope in the Anthropocene can also be projected as a form of being which, deviating from Bloch's (1986) notion of hope as generative possibility, refuses to engage with the all-encompassing, always-generative entangled world.

The chapter is organised into three sections. With the goal of describing the incorporation of entangled ontologies in the discipline of international relations, the first section revisits a genealogy of processual and relational thinking, emphasising that core pioneering works were noticeably situated outside the geographical contours of Europe, where relational ontologies arrived far later. The second part scrutinises critical engagements with entanglement fetishism, organised in a normative spectrum. Seeking to move these criticisms forward, the third section questions the deterministic assumptions behind entangled ontologies, thus redrawing the horizon of hope in the Anthropocene. It examines the possibilities that modes of being reside in the cracks of the relational rubric and, therefore, for reimagining perceived experience as radical openness – namely as an infinite multiplicity of arrays of being, relational and beyond-the-relational – breaking from fetishised exclusionary cuts.

## Revisiting Entanglements in Contemporary Scholarship

The ontological invocation of relations and entanglements as analytical categories enabling hope in the Anthropocene needs to be contextualised in a much broader and older tradition of processual and relational thinking. Whilst continental philosophy only consolidated a process-sensitive philosophy in response to the failures of modernity, as detailed below, in many traditions of thought outside Europe the ontogenesis of being has for centuries been conceived of as processual, bound to limitless emergence, always in relation to other beings and their circumstances. For example, Qin (2009) argues how processes and relations are crucial elements of Chinese political philosophy. In short, Qin argues that, first, processes and beings are inter-constitutive and neither precedes the other; second, processes enable interactions and therefore yield intersubjectivity; third, Confucianism, an ancient Chinese belief system, sees the contours of processes as blurry, therefore taxonomies are not rigid as in the European Kantian tradition. As Qin concludes, if the Western tradition of thought has based its intellectual edifice on rationality, the Chinese philosophical and spiritual tradition has always had relationality at its core (see also Kwang 2006; Nordin et al. 2019). Similarly, Ling (2014) elucidates that Daoist (or Taoist) dialectics, also rooted in Chinese philosophy, are founded on the idea of complementarity, namely that each being retains other beings within, as illustrated by the famous yin/yang symbol (see also Fierke 2019). In addition to these, drawing from ubuntu, an East African and South African philosophy, Ngcoya (2015, 253) conveys how, in this ancient African tradition of thought, 'a person is a person because of/by/through other people'. The author further argues that ubuntu's essential conceptual framework resides in the notions of community, sharing and caring. In other words, true human potential can only be realisable in relations with others, which presumes interdependence and connectedness as a prime condition for being in the world (see also Ramose 1999).

On the American continent, different ancient communities, including the Mayan and the Mapuche, amongst others, share the philosophy behind the Quechua notion of *sumak kawasay*, translated into Spanish as *buen vivir* (good living), which encapsulates the fulfilment of a harmonious life in intimate connection with the planet. Querejazu (Trownsell et al. 2020, 33) describes two essential traits of Andean cosmology. On the one hand, *tinku* is a cosmic function that enables an equilibrium of forces. When beings encounter one another, relations trigger their transformation into something else. On the other hand, *taypi* denotes the '-in between- dimension, process, space of mediation, and connection where difference can encounter and negotiate, where the transformation and the becoming happens' (see also Tickner and Querejazu 2021). As a final example, through dharmic philosophy, which is of key relevance in various Indian religions such as Hinduism and Buddhism, Giorgio Shani and Navnita Behera (2022, 844) suggest that 'Dharma binds human beings to each other and to the universe, including "not only living human beings, but also ancestors, gods, plants, animals, earth, sky, and so on"'. Too often underestimated by continental philosophy, this heterogeneous corpus of theory has articulated for centuries a source of erosion of Eurocentric essentialising ontologies.

In the European continent, the ontology of being was for centuries based on the Aristotelian tradition, for which there is an ultimate being, or substance, that is given (in Latin, *datum*), static, concluded, objectified, 'out-there' waiting to be found by the knowing (human) subject. In the seventeenth century, Baruch Spinoza and Gottfried Leibniz, who were profoundly influenced by the Confucianist intellectual tradition, both pioneered the questioning of this essentialising ontological rubric and systematised the exploration of being as a non-static relational entity. In the second half of the twentieth century, Gilles Deleuze picked up on this processual and relational component in both philosophers. For Spinoza, being enters into composition with other things in existence, therefore things are always composed and decomposed in relation. For Deleuze (1988, 77), in turn, the parts of a whole 'are not themselves individuals; there is no essence of each one, they are defined solely by their exterior determinism, and they always exist as infinities'. In a similar fashion, Deleuze uses Leibnizian thought to suggest that the whole and the parts are not already related, but they encapsulate potential 'derived infinity', namely every possible relational outcome that will be materialised in the world (Deleuze 1993, 46). Thus, in Leibniz, relations are the basis for a primaeval materiality, or what the philosopher refers to as monads:

> relations surge up in a region that [. . .] involves the possibility of creation. [. . .] The whole and the parts are not (and similitude) are not already related, but the original formula of a derived infinity, a sort of intelligible matter for every possible relation: thus the primary terms, without relations themselves, acquire relations by becoming the requisites or the definers of the derived, in other words, the shapers of this material. (Deleuze 1993, 46)

Deleuze insists that relations are themselves types of events, which in turn enter in relation with further events, giving themselves a concrete existence in a continuous spacetime framing.

In the twentieth century, influenced by Auguste Comte, John Dewey, William James and Henri Bergson, all of them renowned processual and relational philosophers, Alfred North Whitehead further explored the relational ontogenesis of being. To Whitehead, who was also knowledgeable about ancient non-European traditions of thought such as Confucianism, relations precede being, what he calls 'actual entities', namely the things that compose the world. In Whitehead, an actual entity is a composite, meaning that 'every actual entity in the universe is constituent of any other actual entity' (Whitehead 1979, 148). More importantly, being is a process of becoming, which is never isolated but constitutively entangled with further processes of becoming, or 'events', considered by Whitehead (2006) the true 'relata', namely the composites of a relation. Essentially, in Whiteheadian philosophy, relationality becomes a tool for moving beyond Cartesian dualism, which bestows an unfounded superiority on human beings on Earth, which over the last three centuries has had devastating effects on the planet as well as generating violent exclusions with respect to race, gender and the canons of beauty. This has been particularly the case insofar as the human subject in the Cartesian imaginary is not just any human subject, but, metaphorically speaking, the image of Leonardo da Vinci's Vitruvian: a white and perfectly proportioned Man (see Mújika Chao 2021).

In historical perspective, the 2000s marked a qualitative split from the majority of previous cross-disciplinary accounts on relations and processes, arguably accelerated by the confluence of two mutually reinforcing processes. First, technoscientific progress in areas such as earth system sciences and science and technology studies led to the definition of a new geological era, the Anthropocene, in which anthropogenic processes have saturated the sphere of human control (see Crutzen and Stoermer 2000). This compelling narrative drew attention to unsettling events, including the potential effects of global warming, desertification and further extreme weather events, alongside the uncertain implications of developments in fields such as neuroscience and artificial intelligence. As a result of these challenges and speculations, the hypothetical finitude of the human being on the planet, far from abstract apocalyptic discourses, has become a strikingly imaginable fate. In other words, the stories about the distinctive, superior and masterful character of the human on Earth seem to be increasingly discredited, and the future appears unquestionably inextricable from broader beyond-the-human phenomena (see Tsing 2015).

Second, the increasing sympathy for claims of the exhaustion and incapacity of the post-positivist paradigm, particularly postmodernism and poststructuralism, as unable to provide analytical tools to enhance for comprehending the repositioning of the human in the Anthropocene era (see Bryant et al.

2011). More precisely, the limits of textual, discursive and semiotic methodological techniques are exposed as insufficient to capture how Anthropocenic processes of transformation are reconfiguring the role of the human on the planet, let alone the relations with its environment. This line of scrutiny states that postpositivist engagements with this quandary, such as the Foucauldian analysis of power relations behind the oppressive effects of modern dualisms and the supremacy of Man (Foucault 1990), overemphasise the deconstruction of one component of the binary, namely 'the social', and underestimate 'the natural', or even the inseparability of both, thus ultimately reproducing and perpetuating a dualist cosmology (see Barad 2007, 145, 209; Kirby 2011, 95; Morton 2013, 4).

In the context of the confluence of these two major processes, recent scholarly interventions have reignited a theoretical and practical response centred on beyond-the-human entanglements and relations (see Latour 2005). In this regard, all beings are rendered vulnerable to the relations that compose them. This erodes and undermines the anthropocentric cosmovision where the human being stands as separated from the world in a position of ontological superiority. According to Manuel DeLanda (2002, 75), becoming is a 'molecular and intensive relation' that operates beyond ontological equilibrium and that is populated by real multiplicities. This framing contributes to the erosion of Cartesian dualism by conceiving of the object, matter and nature as constitutive parts of the subject, mind and culture/the social, respectively. This rethought web of existence invokes a sense of modesty, sensitised with the complex interconnectedness of beings and events in the world, as well as with the forces and (beyond-the-social) power relations that shape the outcomes of this entangled mode of becoming, what Isabelle Stengers (2011) calls non-hierarchical modes of existence.

Events in this new reality, where the mesh of the social entwines with the natural, further undermine the long-assumed separation of the binary, on which the entire project of modernity was illusorily based (see Chakrabarty 2018). More recently, through the notion of relational cosmology, Kurki (2020) projects the universe as an interconnected wholeness. In this account, relations should be thought to precede the existence of the (human and nonhuman) being. In other words, in a Whiteheadian sense, for Kurki, relational emergence outweighs an essential primal being. She therefore ultimately upholds a totalising view on entanglements: 'there is nothing outside of the relations of the universe and our relationality within it' (Kurki 2020, 69). Nonetheless, the acclamation of relations and processes to account for the unfolding of world events, and which fuels discourses of post- or more-than-human hope in the

Anthropocene, has been problematised. The following section scrutinises substantial critical interventions that examine this entangled puzzle.

## Critiques of Entanglement Fetishism

This section highlights a rich and diverse body of literature that expresses caution towards any unrestrained resort to what is hereby conceptualised as entanglement fetishism, namely the celebratory and sweeping claims being made concerning the entangled ontogenesis of being and the totalising projection of a relational, all-too-relational, world. The following lines unpack critical engagements with entangled ontologies by designing a normative spectrum, in which criticisms scrutinise the problematic implications of an assumed entwined genealogy of being.

On one side of the spectrum, authors inclined towards strongly normative and emancipatory stances, including neo-Marxists and poststructuralists, tend to expose the depoliticising slippage that a flat ontological substratum, namely a di-hierarchised and mutually constitutive ontological configuration, might engender. To Erik Swyngedouw and Henrik Ernstson (2018, 19):

> in the transcendence of the nature–society split promised by introducing a human/non-human ontology, the radical otherness upon which relationality is necessarily conditioned is strangely often suspended [. . .] In other words, the move to a relational-materialist ontology sutures things such that the exteriority that undergirds and structures relationality runs the risk of disavowal. To put it simply, the effort to contain and transcend the nature–society split or dualism through ontologies of internal relationality disavows the separation upon which relationality is necessarily constituted.

To these authors, this depoliticising feature of relational ontology ultimately perpetuates the continuity of capitalist society, namely the 'exteriority' or the 'structure' where relationality unfolds. Following a similar theoretical position, Jonathan Joseph (2018, 432) draws from Bourdieu's work on agency and structure to criticise Latour's flat ontological world for eschewing discussion of anything other than practices, thus contributing to 'the reproduction of existing social orders by refusing to examine its conditions of possibility and the hierarchies that lie behind it'. Without disavowing the value of this critique, some uses of entanglements would admittedly escape this depoliticising stigmatisation. For example, whilst Latour's flat ontology might certainly be prone to unleash controversial politics-free ontological frameworks, the use of entanglements in Karen Barad's agential realism is anything but depoliticising, since it articulates sharp epistemological and ethical claims against exclusionary practices of knowledge production (see Barad 2007; Latour 2005).

At the other end of the spectrum, other authors note instead the problematic normative character of entangled ontologies. To Claire Colebrook (2019), this all-out affirmation of a relational world intensifies the normative ethos of being. Being seems reduced to being relational or not being at all. Colebrook picks up on the moralising dimension embedded in the normative ethos of the relational matrix, which is shared by the vast majority of processual traditions of thought: being in relation is morally good. Elucidating on this premise, she argues:

> The norms of simultaneity and relationality have shored up a quite specific Western, European, rationalist morality of the world: humanity is, properly, that which can recognise itself in all the rich cultural variants that make up one interconnected and self-aware whole. To lose that form of humanity would be the end of the world. (Colebrook 2019, 189)

The utter underestimation of the beyond-the-relational, as discussed in detail in the next section, or 'the horror of something that simply is, bearing no relation to anything' (Colebrook 2019, 175), and the privileging of relations and emergent becoming regresses an apparently cutting-edge and radical conversation back to the same exclusionary logics of previous modern ontological cuts, from which social critique in Anthropocene debates so insistently claims to be refraining. Illustrative of what Colebrook discerns, William Connolly (2017, 33) arguably reinforces a moralising humanism intimately associated with the conversation on entanglements. In an attempt to demystify and unveil the weaknesses of sociocentrism and human exceptionalism, the author invites us to forge intellectual and political alliances around what he describes as 'entangled humanism', namely a project 'in which the entanglements include both layered human cultural processes that enter into bumpy relations and nonhuman processes that exhibit periodic unruliness' (Connolly 2017, 68). Thus, Connolly ties the morality underpinning entangled humanism to the human (and humanist) awareness of other modes of experience so as to enhance multispecies appreciation.

At the same end of the spectrum, other authors have exposed the problems involved with the practical materialisation of political projects informed or inspired by relational sensitivities. For example, Chandler, Cudworth and Hobden problematise the normative emphasis in the 'planet politics' debate (see Burke et al. 2016) by comparing proposed policy endeavours such as a hypothetical Coal Elimination Treaty with liberal top-down governmentality.

To Chandler et al., 'Burke et al. spend no time considering what new violences are afforded and enabled in their call for new global governance bodies to "enforce and penalise violence – slow and fast – against nonhuman communities and ecologies" as they seek to legislate for securing the planet against errant humanity' (Chandler et al. 2017, 7). Even more so, the goals of planet politics are 'beyond political negotiation and legal constraints' (Chandler et al. 2017, 8). To be sure, Chandler et al. (2017) state that attempting to control carbon emission through international law 'is unlikely to be effective or to ameliorate planetary inequalities', thus reproducing the colonial violence that sustains capitalism. In this critical view, the implementation of such a treaty would imply mechanisms of enforcement and control which could barely eschew exclusionary practices, thus tripping, once again, over modern ontological cuts. 'Instituting global governance in 'firm and enforceable' ways, as if there were universal solutions that could be imposed from above, is a recipe for authoritarianism and new hierarchies and exclusions', as Chandler et al. (2017, 10) put it. For these critical accounts, entangled sensitivities have gradually transited from being an analytical category to a normative one, namely from trying to decipher the world to norming it. In fact, the authors of the planet politics manifesto, in a later text, defend planet politics' inescapable 'normative commitments' (Fishel et al. 2018, 214).

Interestingly, both sides of the spectrum coincide in their diagnosis. The normative and emancipatory critique, such as that articulated by some neo-Marxists and poststructuralists, stresses the complicity of entangled ontologies with capitalist violence by depoliticising the debate as well as the practices it enables. Criticisms from the opposite end of the spectrum suggest that the type of normative projects designed by relational frameworks can potentially reproduce liberal governmentalities which ultimately underpin capitalist structures of oppression. Seeking to move forward the conversation of hope in the Anthropocene, still very much rooted in the celebration of entanglements, the following section seeks to question an assumption seemingly shared by many of the above critical engagements: the fact that encounters are perpetually generative, thus problematically reducing the ontology of being to a deterministic entangled telos. The possibility of being outside the totalising relational realm seems vastly overlooked.

## From Entanglements to Radical Openness

This section departs from the stance common to both entanglement fetishists and the critical voices unpacked in the previous section: the assumption that entanglements are deterministically generative, as if the ontological preconditions for the relational character of being are already given. Criticisms framed in the above-described normative spectrum are illustrative of this, since the focus of scrutiny is always on the implications unfolding in the aftermath of the ontological encounter. In other words, being is by default drawn into the entangled world, in which, accordingly, an entangled ontological feature becomes the prime condition for possibility.

Seeking to build on and push critical engagements with entanglement fetishists, this chapter suggests the possibility of non-generative encounters, instances where being is outside the all-inclusive relational web. Some beings and events are not given the ontological preconditions to be relational, namely the composites of an entangled world. In addition, the section also discusses the political implications for those subjectivities that reside at the margins of this web of totality. Indeed, most literature tends to conceive of relations and entanglements as deterministically generative, always conducive to further forms of being and becoming. Steven Shaviro (2014, 34) acutely reads this tenet of relations as 'oppressive' and claims that the ultimate metaphysical question is how to escape deterministic and overdetermining relations. With this provocation in mind, this section seeks to delve into the possibility of events where beings or processes refuse to tangle in a generative mode.

Pushing the above-mentioned strands, contemporary literature on relations has also elaborated on what Dillon (2000, 4) calls the 'anteriority of radical relationality', namely a prior stage where the non-relational thrives. Exposing the limits of assumptions on all-out ontological entanglements, Dillon defines the non-relational as:

> [the] utterly intractable, that which resists being drawn into and subsumed by relation albeit it transits all relationality as a disruptive movement that continuously prevents the full realization or final closure of relationality, and thus the misfire that continuously precipitates new life and new meaning. (Dillon 2000, 5)

In a similar line of investigation, Nordin and Smith (2019, 639) add that this non-relational – or what they call 'radical otherness' – can become a recipe against the colonising attempts of a totalising and uniform understanding of the all-encompassing desires of entanglement fetishism. Drawing from Daoist philosophical formulations, the authors hint at radical otherness 'as the unknowable and inaccessible. As a result, this sense of otherness indicates a limit to assimilation, categorization and understanding. The other is constructed as a hard limit for the self, a limit that cannot be broached and assimilated' (2019, 639). Seeking to contribute to these rather intangible dialectics, the argument below examines how the non-relational becomes materially perceptible for certain forms of being whose political subjectivity, and by extension their entanglement with the world, is fundamentally compromised.

An attentive reading of Édouard Glissant's *Poetics of Relation* might offer suggestive elucidations on this form of ontological radicality. Sceptical about the teleological imaginary of a redeemed world in harmonious relation, the French-Caribbean author states that an intense form of being 'cannot bear having any interaction attached to it. Being is self-sufficient. [. . .] Prime elements do not enter into Relation' (Glissant 1997, 161). Drawing on the lived experiences from Caribbean-colonised Creole, Glissant goes on to assert that a sweeping and universalising relational equilibrium subsumes this primal form of existence when non-prime elements, such as race and the violence through which it is reified, are introduced into the equation of being. His notion of 'the opaque' is particularly useful to illustrate this rubric of non-relation. To Glissant, the opaque is what cannot be reduced. Opacity replaces the absorbing concept of unity, refusing the coercive requirements of relation as erasing totality. In Glissant's own words, 'to disindividuate relation is to relate the theory to the lived experience of every form of humanity in its singularity' (1997, 195).

It follows from Glissant's stance that, first, the ontological preconditions for being entangled with the world are not deterministically given. Different to assumptions behind fetishised positions as well as most of its detractors, being is not already drawn into the cosmos. Second, the underlying condition for survival of a prime, irreducible form of being, which is alien to the world as relational and colonising totality, can only be preserved through a form of political resistance driven by non-engagement, refusal and even withdrawal. Kennan Ferguson (2021, ix) opens up a refreshing conversation about absolute refusal as a valid mode of thinking and acting. The author questions our form of being as conducive to a *sine qua non* involvement with reality. Ferguson's position refuses being as communal experience and collective purpose. In contrast to the popular 'we are all in this together', often invoked in Anthropocene conversations pointing to humanity as a whole, he insists that certain historical legacies preclude certain kinds of people from engaging in the political space (Ferguson 2021, xvi). For example, apologetic symbolism offered by former colonial powers to descendants of enslaved people clashes with the process of reconciliation, which requires subsumption, 'get over it'. Thus, non-engagement, non-reconciliation becomes the political goal (Ferguson 2021, xvii).

In a similar vein, Frank B. Wilderson III (2021, 94) uses the condition of Blackness to illustrate the impossibility of a liveable relational ontogenesis of the Black reality. This author compares the plight of the Indians in the Americas, which he defines as a reciprocal dynamic acknowledged between degraded humanity (Indians) and exalted humanity (white settlers), with the struggle between Blackness and the world. To Wilderson, 'social death', namely a sense of ontological erasure exemplified in the violence perpetrated in the slavery event, manifests the absence of generativity in the encounter: the slave is deprived of modern-like political subjectivity and therefore of the potential for its engagement with the world. Whilst Wilderson's relation-deprived ontology of the slave can be read as deriving from an annulling process of colonial dispossession, Glissant's ontology can be read to unearth a bare, primal, non-relational form of being. However, both coincide, in that the only possibility for being engaged with modern relational totality is through the logics of violence, oppression and subjugation. To these authors, these aesthetic actualities are ontologically excluded from a quasi-glorified, delusional, all-encompassing entangled form of being. More importantly, in these ontological margins, non-engagement, refusal and withdrawal become a form of political resistance that preserves this mode of being, otherwise absorbed by the erasing force of relational totality (see also Colebrook 2022).

Unexpectedly, the retreat from the political turns out to be unambiguously politicising: Culp (2021, 111) suggests that 'the way through is not to pity the nihilism of nonbeing in a move to redeem them; rather, the non- of the nonbeing and its survival is an ultimatum that invited an insurrection against the world'. In Anthropocene terminology, these insights insist against saving the world and instead suggest ways to end it, as Culp notes. Rather than looking for solutions towards a realisable entangled world, this approach declares the insufficiency of everything, which should not be confused with political inaction. The destructive character embodied in this sense of being without being in relation, the utmost refusal, the author continues, disrupts the long game of the political projects of coalitions, for example the accumulating of liberal democratic rights as if this would eventually conclude in human liberation.

This line of argumentation has been unfolded in this section as an invitation to move critical debates beyond the contours of deterministic and fetishised ontological entanglements. Anticipating the potential of this critical undertaking, Chipato and Chandler (2023, 166) have recently put into question the assumptions behind an all-in cosmological relationality, as if reality is the composite of a multiplicity of worlds or universes, often referred to as the pluriverse (see Trownsell et al. 2020). As Chipato and Chandler observe, this strand of literature overlooks and neglects the genesis of exclusionary ontologies, that is, 'the preconditions for a world of "many worlds" in which there are plural modes of becoming human', thus amplifying the risks of reproducing rather than eroding colonial violence. Crucially, to these authors, 'the escape into a vital, relational, creative world of complexity, and the search for new ways to become with nature helps to obscure how we got to the world that we have' (Chipato and Chandler 2023, 164). Rather than the infallible and continuous generativity embedded in most literature on the pluriverse, Chipato and Chandler project a cosmos where being and its presumed deterministic relational condition of subject fade.

Thus, through the exposure of the limits of the entangled narrative, this chapter builds on the attempts at reimagining perceived experience as radical openness, resonating with what Glissant (1997, 8) calls the abyssal 'unknown'. A post-Cartesian ontological emancipation cannot be attained by reducing ontology to an entangled yoke, or worse, to the relational human experience of the world, for this would just reproduce the exclusionary violence of modern logics and practices (for a discussion see also Waldow, this volume). Instead, an honest acknowledgement of the ontological vulnerability of being, so often claimed by discourses on hope in the Anthropocene, would eschew such an overdetermining hurdle and be open to the possibility of being beyond this all-embracing entwined condition. The leap from entanglement fetishism to radical openness entails a view of hope in the Anthropocene as the possibility for unlimited arrangements, which hinge on unpredictable forces that enable all generative collisions but also incompossibilities, thus unearthing thought and practice towards hope beyond the Blochian idea of hope as generative possibility (Bloch 1986).

Far from firm arguments, this exploratory chapter has invited critical reflection upon the invocation of harmonious relationality as a form of projecting hopeful futures in the Anthropocene age. Whilst most popular theoretical and policy accounts suggest that hopeful and liveable futures require materialising durable entanglements, this chapter has sought to speculate on the possibility of being beyond the determinism embedded both in the full-blown entanglement narrative as well as in most critical engagements with it. Building on these critical voices, the chapter attempts to move the conversation forward by interrogating the seemingly inescapable generativity of relations: the assumption that entanglements deterministically precipitate further beings and events. With the goal of offering an alternative notion of hope in the Anthropocene, the text has explored contemporary literature to discuss ontological possibilities for being beyond the determinism of entanglements as well as the political implications of being in the margins of the all-embracing relational rubric. Specifically, the chapter has sought to push perhaps unintuitive notions such as radical relationality, namely the anteriority to relationality, by stressing the material dimension of beyond-the-relational political subjectivities whose engagement with an entangled world can only be reified through the logics of violence, oppression and subjugation, and for which refusal, non-engagement and withdrawal become a mode of resistance and survival.

In all, the intent has been a broader goal, namely the reimagining of perceived experience in our Anthropocene times in terms of radical openness. By reducing the ontology of being to a deterministic entangled ontogenesis, emancipatory claims about a plural and all-inclusive hopeful future in the Anthropocene are utterly compromised. As a word of caution, the text hints that interventions around entanglements might offer promising ontological and political paths to the poetics of hope in the Anthropocene as long as these are sensitive to the infinite multiplicity of possibilities for being, relational and beyond-the-relational, breaking from a seemingly fetishised and exclusionary form of hope where being in relation is deterministically set as the prime condition for possibility.

## References

Barad, K. 2007. *Meeting the Universe Halfway: Quantum Physics and the Entanglement of Matter and Meaning*. Durham, NC: Duke University Press.

Bloch, E. 1986. *The Principle of Hope*, vol. 1. Cambridge, MA: MIT Press.

Bryant, L. R., Srnicek, N. and Harman, G. 2011. *The Speculative Turn: Continental Materialism and Realism*. RE.press.

Burke, A., Fishel, S., Mitchell, A., Dalby, S. and Levine, D. J. 2016. Planet politics: A manifesto from the end of IR. *Millennium: Journal of International Studies* 44(3): 499–523.

Castro Pereira, J. and Saramago, A. (eds). 2020. *Non-Human Nature in World Politics: Theory and Practice*. Springer Link.

Chakrabarty, D. 2018. Planetary crises and the difficulty of being modern. *Journal of International Studies* 46(3): 259–82.

Chandler, D., Cudworth, E. and Hobden, S. 2017. Anthropocene, Capitalocene and liberal cosmopolitan IR: A response to Burke et al.'s 'Planet Politics'. *Millennium: Journal of International Studies* 46(2): 190–208.

Chipato, F. and Chandler, D. 2023. The Black horizon: Alterity and ontology in the Anthropocene. *Global Society* 37(2): 157–75.

Colebrook, C. 2019. A cut in relationality: Art at the end of the world. *Angelaki* 24(3): 175–95.

Colebrook, C. 2022. Deleuze after Afro-pessimism. In C. Daigle and T. H. McDonald (eds), *From Deleuze and Guattari to Posthumanism: Philosophies of Immanence*. London: Bloomsbury Academic.

Connolly, W. 2017. *Facing the Planetary: Entangled Humanism and the Politics of Swarming*. Durham, NC: Duke University Press.

Crutzen, P. J. and Stoermer, E. F. 2000. The 'Anthropocene'. *Global Change Newsletter* 41: 17–18.

Culp, A. 2021. Afro-pessimism and non-philosophy at the zero point of subjectivity, history and aesthetics. In K. Ferguson (ed.), *The Big No*. Minneapolis: University of Minnesota Press, 105–35.

DeLanda, M. 2002. *Intensive Science and Virtual Philosophy*. London: Continuum.

Deleuze, G. 1988. *Spinoza: Practical Philosophy*. San Fransisco, CA: City Lights Books.

Deleuze, G. 1993. *The Fold: Leibniz and the Baroque*. Minneapolis: University of Minnesota Press.

Dillon, M. 2000. Poststructuralism, complexity and poetics. *Theory, Culture & Society* 17(5): 1–26.

Ferguson, K. (ed.). 2021. *The Big No*. Minneapolis: University of Minnesota Press.

Fierke, K. M. 2019. Contraria sunt complementa: Global entanglement and the constitution of difference. *International Studies Review* 21(1): 146–69.

Fishel, S., Burke, A., Mitchell, A., Dalby, S. and Levine, D. 2018. Defending planet politics. *Journal of International Studies* 46(2): 209–19.

Foucault, M. 1990. *The History of Sexuality*, vol. 1: *An Introduction*. London: Vintage.

Glissant, É. 1997. *Poetics of Relation*. Ann Arbor: University of Michigan Press.

Haraway, D. 2016. *Staying with the Trouble: Making Kin in the Chthulucene*. Durham, NC: Duke University Press.

Joseph, J. 2018. Beyond relationalism in peacebuilding. *Journal of Intervention and Statebuilding* 12(3): 425–34.

Kirby, V. 2011. *Quantum Anthropologies: Life at Large*. Durham, NC: Duke University Press.

Kurki, M. 2020. *International Relations in a Relational Universe*. Oxford: Oxford University Press.

Kwang, K. 2006. *Confucian Relationalism: Cultural Reflection and Theoretical Construction*. Beijing: Beijing University Press.

Latour, B. 2005. *Reassembling the Social: An Introduction to Actor-Network-Theory*. Oxford: Oxford University Press.

Ling, L. H. M. 2014. *Imagining World Politics: Sihar and Shenya, A Fable for Our Times*. Abingdon: Routledge.

Morton, T. 2013. *Hyperobjects*. Minneapolis: University of Minnesota Press.

Mújika Chao, I. 2021. El género del fin del mundo: aportes de la investigación feminista por la paz ante el mantropoceno. *Revista de Estudios en Seguridad Internacional* 7(1): 45–60.

Ngcoya, M. 2015. Ubuntu: Toward an emancipatory cosmopolitanism? *International Political Sociology* 9(3): 248–62.

Nordin, A. H. M. and Smith, G. M. 2019. Relating self and other in Chinese and Western thought. *Cambridge Review of International Affairs* 32(5): 636–53.

Nordin, A. H. M., Smith, G. M., Bunskoek, R., Huang, C., Hwang, Y., Jackson, P. T., Kavalski, E., Ling, L. H. M., Martindale, L., Nakamura, M., Nexon, D., Premack, L., Qin, Y., Shih, C., Tyfield, D., Williams, E. and Zalewski, M. 2019. Towards global relational theorizing: A dialogue between Sinophone and Anglophone scholarship on relationalism. *Cambridge Review of International Affairs* 32(5): 570–81.

Qin, Y. 2009. Relationality and processual construction: Bringing Chinese ideas into international relations theory. *Social Sciences in China* 30(4): 5–20.

Ramose, M. B. 1999. *African Philosophy through Ubuntu*. Mond Books.

Shani, G. and Chadha Behera, N. 2022. Provincialising international relations through a reading of *dharma*. *Review of International Studies* 48(5): 837–56.

Shaviro, S. 2014. *The Universe of Things: On Speculative Realism*. Minneapolis: University of Minnesota Press.

Stengers, I. 2011. *Cosmopolitics II*. Minneapolis: University of Minnesota Press.

Swyngedouw, E. and Ernstson, H. 2018. Interrupting the Anthropo-obScene: Immuno-biopolitics and depoliticizing ontologies in the Anthropocene. *Theory, Culture & Society* 35(6): 3–30.

Tickner, A. B. and Querejazu, A. 2021. Weaving worlds: Cosmopraxis as relational sensibility. *International Studies Review* 23(2): 391–408.

Torrent, I. 2021. *Entangled Peace: UN Peacebuilding and the Limits of a Relational World*. Lanham, MD: Rowman & Littlefield.

Trownsell, T. A., Tickner, A. B., Querejazu, A., Reddekop, J., Shani, G., Shimizu, K., Behera, N. C. and Arian, A. 2020. Differing about difference: Relational IR from around the world. *International Studies Perspectives* 22(1): 1–40.

Tsing, A. 2015. *The Mushroom at the End of the World: On the Possibility of Life in Capitalist Ruins*. Princeton, NJ: Princeton University Press.

Tucker, K. 2018. Unraveling coloniality in international relations: Knowledge, relationality, and strategies for engagement. *International Political Sociology* 12(3): 215–32.

Wendt, A. 2015. *Quantum Mind and Social Science: Unifying Physical and Social Ontology*. Cambridge: Cambridge University Press.

Whitehead, A. 1979. *Process and Reality*. Free Press.

Whitehead, A. 2006. *The Concept of Nature*. Project Gutenberg.

Wilderson III, F. B. 2021. Without priors. In K. Ferguson (ed.), *The Big No*. Minneapolis: University of Minnesota Press, 85–103.

Zanotti, L. 2019. *Ontological Entanglements, Agency and Ethics in International Relations: Exploring the Crossroads*. Abingdon: Routledge.

# 15

# Hope as a Theopolitical Virtue: Eschatology and End-of-Time Politics

*Vassilios Paipais*

**Introduction**

In the grip of an apocalyptic mood in the aftermath of the Second World War, Martin Wight famously quipped that 'hope is not a political virtue; it is a theological virtue' (Wight 1948a, 3). Such an aphorism may have made sense in the context of the atomic age and in the wake of the apocalyptic violence unleashed by twentieth-century totalitarianisms that Wight (1955), with Eric Voegelin, saw as the 'political religions' of our time. Wight (1948b) was prudent enough to issue a warning against both the dangers of the apocalyptic imaginary in world politics (the violent and premature introduction of salvation in history) and its negative mirror image, the dangerous sacralisation of sovereign power (performing the sacred duty of the *katechon*, the 'withholder' or 'restrainer' of apocalyptic chaos) that Carl Schmitt's order-oriented decisionism sanctions (Schmitt 2003 [1950]). Still, these options (either apocalypse/disorder or its katechontic containment) do not exhaust the breadth, fecundity and promise of eschatological thinking, nor do they appreciate the spiritual, prophetic, justice-oriented dimensions of religious hope.

Indeed, such considerations may have assumed a renewed importance and urgency in the age of the so-called Anthropocene, especially since the Anthropocene does not only inaugurate the 'real' end of the world against which the cherished illusions of modernity, such as reason, humanity, progress, have faltered but also offers an invitation to rekindle our imagination in search of alternative conceptualisation of critique, agency and politics. In this context, eschatology as a traditional theological discourse about the end of times cannot remain unaffected but needs to consider the nihilist context within which investigations of hope in a (post-)apocalyptic environment take place. If the end of the world has already taken (or is continuously taking) place, what does it mean to live in hope in such a context and how can eschatological thinking save us from turning the apocalypse into another technique for the management of human existential anxiety and insecurity?

This chapter takes this challenge seriously by exploring the meaning of hope as a theopolitical virtue in a nihilist era, inaugurated by Nietzsche's proclamation of the 'death of God' and Heidegger's critique of Western metaphysics. Within the nihilist horizon, hope as simultaneously a theological and a political virtue is envisioned as equidistant both from arguments that favour a sanitised separation of eschatology from politics *and* from those that tend to recruit it in the service of earthly, political or technoscientific, utopias. In making the case for eschatological hope as a disruptive sensibility in a disenchanted world 'where gods have departed', it is crucial to distinguish between *anti*-political eschatologies driven by an apocalyptic imaginary that reflects a dualistic, otherworldly, post-historical rendition of the eschaton, and *counter*-political eschatologies advancing a this-worldly, historically engaged, messianic vision of the eschatological impulse as a force of undoing.

The analysis begins by tracing the development of the subversive impulse in the modern revival of eschatology that raised the question of the relationship between divine revelation and human agency, eschatology and history, in the context of the modern crisis of nihilism. It then examines how religious hope, in the dialectics of Jewish messianic and Christian Trinitarian theocratic anarchism, becomes not a call for an other-worldly or vacuous negativism, but a counter-political critique of the mystification of power and the logic of mastery in history. Contra Wight, hope as theopolitical virtue becomes the terrain upon which theology and nihilism enter a zone of indistinction, thus opening the political to another use that challenges the sacred doctrines of both traditional theology and secular humanism as well as some of the most emblematic technological eschatologies of the Anthropocene embodied in contemporary post- or transhumanist futurism.

## History, Eschatology and Political Theology

Religious hope, in the form of apocalyptic eschatology, traditionally invokes images of destruction of this world in favour of a world to come. Politically, it is linked to the radical idea of the rejection of a corrupt world given to sin, evil and suffering for a world of redemption and eternal happiness. Yet, as scholar of eschatology Judith Wolfe reminds us (2022, 334), the recovery of the apocalyptic in late nineteenth- and early twentieth-century biblical scholarship was not initially motivated by the desire to retrieve its radical spirit. Instead, it was the work of the History of Religion School that arose from a commitment to historicism (see Paipais 2018, 1016). Its aim was to uncover the specific spatio-temporal circumstances, actors and influences on Christianity in its historical development and, in the process, separate what was perceived as the primitive apocalyptic mindset of early Christianity, which preached an imminent end of history that did not arrive (Schweitzer 1968 [1893]; Weiss 1971 [1892]), from a demythologised 'enlightened Christendom' that, since Augustine and Aquinas, had interiorised – and, consequently, pacified – the eschatological impulse (Phillips 2015, 277–9).

A pivotal role in this development was played by the New Testament scholar Franz Overbeck (1837–1905) who formulated the 'disjunction thesis'. In his provocative book *Über die Christlichkeit unserer heutigen Theologie* (1873, 2nd edition 1903), Overbeck argued that an absolute gap separated the apocalyptic radicalism of primitive Christianity, rejecting any hope of salvation within this world, and the subsequent institutionalisation of the Christian faith that reconciled the church with history and secular powers. Christianity's 'historicisation' was perceived as an act of betrayal, self-deception and accommodation with the powers of this world, diluting faith through reason and promising salvation within history. The implications of Overbeck's radical claim divided his readers. Philosophical radicals, such as Friedrich Nietzsche and Martin Heidegger, drew the conclusion that, in light of the refutation of imminent eschatological expectation, the Christian faith had simply become untenable. The young Swiss theologian Karl Barth, by contrast, saw in Overbeck's antinomian reading the condition of possibility for the radicalisation of Christian faith, conceived as the expectation for the inbreaking of God into the lives of the faithful against all worldly certainties and religious complacencies.

Both interpretations, however – philosophical and theological – converged on the mutual acknowledgement that, in the wake of the ruins of the First World War, history could no longer be seen as heading to some Hegelian fulfilment. If there was to be eschatological judgement, or fulfilment of any kind, it would probably resemble the countercultural apocalypticism of the Bible, rather than the secular theodicies of humanist optimism or the self-righteous moralism of liberal Protestantism. Early twentieth-century eschatology, in other words, came to represent not a force that consummates but rather one that interrupts the flow of history, either through revolution or by withdrawing from history entirely, giving rise to a dualism that opposed history to eschatology, politics to redemptive faith.

Indeed, the rise of the interwar 'theology of crisis', with Barth as its leading figure, can be explained as an expression of this radical, countercultural spirit of dualism that emerged out of the frustration caused by the horrors of the

First World War (see Hotam 2007; Lilla 2007). For this reason, Barth's dialectical 'theology of crisis' has often been described as a modern revival of Gnosticism with its radical spirit of dualism between an evil world and a transcendent Deus Absconditus absolved of the world's malevolence. Adolf von Harnack, for example, dismissed Barth's theology as 'Gnostic occultism', arguing that crisis theology's absolute emphasis on divine transcendence implied a Gnostic denigration of the immanent world (Styfhals 2019, 32). In his famous interwar dissertation on Gnosticism, Hans Jonas (1992 [1934]) similarly points to a connection between the Gnostic teachings of Marcion of Sinope – a Christian Gnostic who rejected the 'evil' God of the Old Testament – and interwar theology, arguing that almost the entirety of contemporary theology was saturated with the spirit of Marcionism (see also Lazier 2008, 33; Lilla 2007).

Although it is commonly agreed that the rise of neo-Gnostic antinomianism (from the Greek anti-*nomos*, 'against the law') in the interwar period reflects a subversive tendency shared by both Christian and Jewish radical thinkers, careful distinctions need to be drawn. Gnosticism's negative antinomianism conceives the world itself as evil, the product of an inferior god (literally or metaphorically). Gnosticism then wages war not against the present state of the world but against the world as such. While the spirit of antinomianism is critical and disruptive of the sedimented structures of this world (the status quo), it is not Gnostic per se as it does not turn against the world as such but only against the logic of this world – what Walter Benjamin (2007a, 277–300) in his 'Critique of Violence' described as the origin of the law in retributive violence. A common misunderstanding here is that such an attitude denotes an aversion to historical action or some anti-political, purist retreat to passivity, quietism, detachment or resignation. That this is not necessarily, or perhaps primarily, the case (and why this makes modern eschatology a counter-political rather than an anti-political discourse) is foregrounded in the work of a pioneer Jewish political theologian whose work inspired many of the antinomian radicals of the period, Martin Buber (1878–1965).

## Messianic Hope and Human Agency

Buber defends exactly the possibility of politico-religious hope as an antinomian, disruptive discourse that does not reject the world as evil but reorients human agency in its pursuit of justice in the context of an alienated reality. In his 1932 book *Kingship of God*, Buber (1990) proposed that the ancient Israelite political theology is grounded on a unique historical mission to create an alternative political system based on justice rather than pure power. The warning to Israelites to 'remember that it is the Lord your God who gives you power' (Deut. 8:18) served as an alternative model to the realpolitik of the ancient world whereby the earthly power of pharaohs and emperors was its own source of divine authority. The idea that the tyranny of kings could be replaced by the reign of God, whose will could only be reflected through the righteous rather than the powerful, was part of an early form of 'anarchistic theocracy' that would later filter into the Kabbalistic messianism and religious nihilism of Gershom Scholem and Walter Benjamin (see also Vatter 2021).

Buber was among the first of a series of nineteenth- and twentieth-century literati of the Jewish Central European intelligentsia that connected Jewish mystical messianism with this ancient anarchic theocratic ideal (Löwy 1992). In his effort to differentiate between the redemptive, dangerous hope of the Hebrew prophets and the eschatological hope or inner messianism of the rabbinic tradition that put emphasis on waiting for God to redeem the world at the end of days, Buber erected his model of God's kingship on the idea of a dialogic interaction between the divine and the human in history. As opposed to the fatalism that the spiritualisation/internalisation of messianism entailed within the rabbinic and early Christian traditions, Buber insisted that messianic hope rested on the dialectics of the human reaching out towards the divine in search of recognition and dialogue. Such a model of human–divine dialogic interaction did not constitute another ritualistic form of worship but rather inaugurated a universalistic faith eventually uniting all God's children in peace and justice. Instead of assigning sacral authority to ecclesiastical and social hierarchies represented by what Buber called the *priestly principle*, the prophets created an alternative devotional model of God's an-archical ('no rule') sovereignty (see also Lerner 2015).

Buber, in other words, laid the foundations for what would later be understood as the antinomian tendency in Jewish messianism, the idea that true prophetic faith oriented to justice and to living the life of God in this world challenges legal and power structures upon which not only kingly rule rests but also the sacral authority of priests and ecclesiastical institutions. Scholem would later insist, in his famous essay 'The Messianic Idea in Judaism', that the antinomian or 'heretical' impulse is constitutive of Jewish messianism (Scholem 1971). However, the question is not whether a countercultural, revolutionary spirit punctuates the prophetic messianism of the Jewish faith, but what exactly this antinomianism entails and how different it is from Gnostic other-worldliness. Otherwise put, is it a nihilistic rejection of the law or, as Agamben (2005) and Vatter (2021) argue, in their own respective ways, the true fulfilment of the law (*Aufhebung*) in love or, equally, in the life of God, as expressed in the Aristotelian/Philonian idea of God as the Living Law (νόμος ἔμψυχος)?

The question of the nature of messianic antinomianism and its difference from apocalypticism was from the outset linked by Buber to the role of human agency in history. Akin to the polarity in his earlier writings between religion and religiosity, the messianic for Buber leans heavily on the possibility of spiritual agency and praxis in the world, whereas the apocalyptic embodies these qualities of religious faith that lead to an overdependence on God as an external force acting within history from the outside. In other words, while messianic yearnings rely on a more interactive understanding of the relationship between divine and human agency, apocalypticism gives credence to transcendental forces 'breaking in upon history', in Scholem's (1971, 10) formidable phrase.

## The Dialectics of Messianic Nihilism

Whereas Buber was keener to emphasise the redemptive potential of messianic fervour within history, his students Gershom Scholem and Franz Rosenzweig were a lot more ambivalent about the vitalistic core of all messianic movements, an almost 'demonic' anarchical power seeking 'redemption through sin' and hope through the destructive joys of its own negation (Scholem 1971, 78ff.). The antinomian fervour of messianism could subvert sedimented structures and rigid social hierarchies. However, when linked to politics, it could also become violent and destructive. Arguably, the antinomian spirit of Gnostic revolt in apocalyptic messianism mirrors the arbitrariness of the Schmittian state of exception (Schmitt 2006 [1922]) to such an extent that, in Scholem's (1971, 1–36) interpretation of the tension between restorative, utopian and revolutionary impulses of Jewish messianic narratives, messianic antinomianism tends to ambivalently oscillate between violent imagery and political quiescence. This volatile vitalism at the core of the messianic impulse made it more difficult to clearly separate the history of prophetic messianism as God's anarchic sovereignty on earth from apocalyptic nihilism, which manifests in two antithetical but equally problematic ways, either as the 'redemptive sin' of destructive revolutionary zeal or as detachment from a corrupt world and perpetual deferral of real social transformation.

In Scholem's case, this ambivalence at the heart of messianic vitalism was a result of the structural, albeit always antinomian and subversive, legacy of Gnosticism in Jewish messianism traceable both in the ancient Jewish mystical tradition and the modern mystical heresies of Kabbalah and Sabbateanism. Making no distinction between modern and Gnostic forms of nihilism, Scholem interpreted Gnosticism as a constant structural temptation in religious thought – both Christian or Jewish – enacted in different times, in both ancient and modern contexts. In both cases, he was interested in the dialectics between a world-negating focus on redemptive hope, close to the Kafkaesque idea of 'hope for the hopeless', and the continuation of an unredeemed life in the present. To his mind, Gnostic antinomianism ('transcendence breaking in upon history') showed very clearly how the way one inclines to the questions of transcendence and salvation conditions the way one behaves in the immanent world and relates to profane history, even if the Gnostic concept of transcendence implies a radical rejection of immanence and history (see Styfhals 2019, 136–46).

In fact, Scholem reversed the traditional Gnostic valorisation of transcendence by putting its antinomian spirit of revolt in the service of immanence. In a world where God is absent, or even pure nothingness, the most pressing task is not how to reach transcendence but rather how to continue living in this hopeless world; how to affirm this world despite, or even through, its fundamental nihilism and meaninglessness (Styfhals 2019, 137–8). In that sense, he proposed a revaluation of life in this world in a way that read its immanent logic as a manifestation of divine nothingness. Such a conception of religious nihilism stands as the inversion of all things earthly and, paradoxically, as a nihilistic affirmation of this world. His historical analyses of Sabbatean and Gnostic subversion demonstrated the attraction and ambivalence of such a nihilistic inversion that employed a complex and sublimated antinomianism as the inverse of a lost transcendence.

This motif of nihilistic inversion also returned in the work of Scholem's close friend, Walter Benjamin (2007a, 312–13), notably in his 'Theological-Political Fragment'. Benjamin's outlook in the 'Theological-Political Fragment' is often described as radically dualistic, with aphorisms like 'nothing historical can relate itself on its own account to anything Messianic' often invoked to make the case (2007a, 312). This could indeed come across as a Gnostic absolute division between immanence and transcendence: God is completely absent from world history, which is dismissed as finite, inferior and radically meaningless. Yet, Benjamin, like Scholem, was not concerned with affirming the priority or purity of the transcendent but entertained a dialectical interest in immanence. The absolute separation of the profane from the messianic is posited as a way of determining the meaning of the former as the dialectical inversion of the messianic: 'one arrow points to the goal toward which the profane dynamic acts, and another marks the direction of Messianic intensity' (2007a, 312). While the profane has its own autonomous logic, it can only be perceived as a complete inversion of the logic of the messianic which, paradoxically, is the only one that can consummate it. This opposition does not devalue immanent existence, but rather intimates that its meaning can arise paradoxically only in opposition to the messianic posited as absent or nihilistically constituted (see also Paipais 2018,

1027). Benjamin here suggests that the revalorisation of the profane does not depend on a value that has to be added to or superimposed on it, but can exist only in this very meaninglessness and opposition itself. Meaning, for Benjamin, arises in the essential transience of the profane, true hope in hopelessness.

Benjamin conceived history itself as antinomian and messianic by virtue of its essential transience. As he emphasised in his 'Theses on the Philosophy of History', history itself has 'a *weak* Messianic power', as it 'carries with it a temporal index by which it is referred to redemption' (Benjamin 2007b, 254). If this is the case, the messianic is at work in the present itself and constitutes meaning within the profane. This would have been impossible for Scholem, whose experience of profane meaning is merely one of nothingness and radical divine absence. In line with Scholem, Benjamin was interested in the dialectical implications of a radical other-worldliness for the meaning of *this* world, but, unlike Scholem, Benjamin's interest in other-worldliness itself was subordinated to a focus on the profane. Benjamin radicalised Scholem's dialectics to such an extent that he arguably overcame Scholem's Gnostic frame of reference. While Scholem's dialectics oscillated between a focus on the absent God and a return to the world, Benjamin's messianic nihilism was concerned with saving the meaning and autonomy of the world against Gnostic transcendence (for a similar reading, see Bielik-Robson 2019).

Ultimately, Benjamin's nihilism finds hope in the abyssal depths of the annihilation of all false hopes that modern politics invests in only to pile up one catastrophe on top of another, to paraphrase his Angel of History parable. And yet, his world nihilism remains a fully political position. Messianic nihilism calls for a retreat from politics, but only if the latter is conceived as the playground of the masters and victors of history. As such, it raises the possibility of another politics qua refusal to conform to the historical mainstream of linear progress as the justification of political success in history.[1] The messianic involvement with worldly politics is not one of gradual change or improvement of the political status quo but rather one of withdrawal. World politics as nihilism undermines the aura that surrounds the *auctoritas* of political power.

## Theocracy in Jewish Messianism and Christian Trinitarianism

The question of meaning, agency and hope in an unredeemed, nihilist world was also an important leitmotif in the thought of Jacob Taubes (2009 [1947]). Taubes approached this question almost always from the point of view of what he called apocalypticism. Like Scholem and Benjamin, Taubes was interested in how the realisation that there is an end to time influences the way we live in history. Not unlike Gnosticism and messianism, apocalypticism radically put the spiritual meaning of the world in jeopardy. If God is supposed to destroy the world at the end of time, the history of the world has to be rejected as transitory, finite and even radically evil. Nonetheless, as history has not yet come to an end, the apocalypticist inevitably must decide how to live in, and make sense of, history as a transitory period. Again, the passivity of resignation and anticipation cannot be a viable option. Although the apocalypticist rejects profane history and politics, she cannot escape her involvement in them. In other words, she cannot escape the simple fact that she lives in history and is part of a political community.

In *The Political Theology of Paul*, Taubes (2004) argued that these are also the questions that the apostle Paul had grappled with. In Taubes' view, Paul's project is political theology: he established a political community based on the theological conviction that there is an end to time and that he lived in a transitory period. In this politico-theological spirit, Paul's aim was the 'establishment and legitimation of a new people of God' (Taubes 2004, 28). Paul's apocalyptic community is what we still know today as the church. For obvious reasons, the church has tried to suppress its apocalyptic roots as much as possible. Since Paul believed that the end of the world was near, the political legitimacy of his community was essentially transitory. The community itself had no absolute legitimacy because it was to be abolished at the end of time together with all worldly and political affairs. According to Taubes, Paul's apocalyptic worldview delegitimised political order as such (more explicitly, the historical form political order had assumed in his time, the Roman Empire, but also its counter-model, the particularistic Jewish ethnic community). Not unlike its more radical Gnostic variants, Paul's church prepared and even strove for its own abolishment at the end of time.

Nonetheless, it would be too hasty to presume that, for Taubes, Paul's political community is premised on the apocalyptic rejection of all politics, as Terpstra and de Wit (2000) suggest. Taubes does not endorse an anti-political quietism, nor does he necessarily subscribe to a politics of revolutionary transformation whereby the sovereign is replaced by the people as the incarnation of divine sovereignty, a politics that resurrects the sovereign imperative in inversed form. Rather, much like Benjamin, Taubes (2004) maintains that this antinomian refusal to act was nonetheless a political act, however empty, nihilistic and antipolitical it may seem. The messianic rejection of the powers of this world remained ultimately a political position on which a purely horizontal, 'free of rule [*herrschaftsfrei*]' community of 'the people of God' could be established. In this regard, apocalypticism had a political theology, albeit necessarily a negative one (*spiritual* as opposed to *worldly*, in Taubes' parlance): theology no longer legitimised a certain political order, as in Carl Schmitt, by showing how it represented the divine, but delegitimised sovereign political order as such (and the idea that the divine can be represented on Earth). This negation of the legitimacy of every earthly political order becomes the cornerstone of an anarcho-theocratic apocalyptic politics.

Interestingly, as Christoph Schmidt (2007, 238–9) points out, Taubes' presentation of Paul's political theology manifests 'more than just a deep affinity with the reconstruction of Paul's Epistle to the Romans by the 1929 convert to Catholicism Erik Peterson'. For Schmidt, Taubes' book on Paul reveals 'the quintessence of his constant vacillation between Jewish and Christian traditions', while his 1947 doctoral thesis 'Occidental Eschatology' 'owes far more than is fitting to the Jesuit Hans Urs von Balthasar's *Apocalypse of the German Soul*'. Taubes' deep affinity with the Catholic Church fathers of the twentieth century that renewed Catholic theology and inaugurated the movement of *nouvelle théologie* is a testament to the subterranean connections between the modern revivals of Jewish political theology and Christian Trinitarianism – the doctrine that the triune Christian God reveals the mystery of uncreated reality, the loving *synoikēsis* (= mutual indwelling) and *perichoresis* (= circumincession) of three divine person in a single divine essence.² The common programme here is not simply the rejection of the idea that any earthly political order can be modelled on, or reflect, divine rule or justice (because no earthly rule can *represent* the transcendence of Yahweh or the triune God), but also that Jewish/Pauline antinomian messianism and Christian Trinitarianism authorise a public or political form of conduct one could describe as theocratic an-*archy* ('no rule') (see also Paipais 2022).

The Catholic convert theologian Erik Peterson, a major influence on the *nouvelle théologiens*, is famous for stressing the political significance of theocratic eschatology encompassed in Christian Trinitarianism. As György Gereby has superbly shown, Peterson's great discovery in his 'Eis Theos' doctoral dissertation was that the congregation of Christian believers (*ekklesia*) was a political community par excellence constituted through an acclamation of faith that was both religious and political without conferring any sacrality on political or ecclesiastical power (Gereby 2008; see also Hollerich 2011). Peterson's knowledge of the Cappadocian fathers of the church (heirs to the Philonian tradition of reconciling Judaism and Hellenism)³ demonstrates that, once orthodox Trinitarian theology was fully developed in the fourth century AD, any possibility for the sacralisation of power through Trinitarian analogies becomes impossible due to the orthodox dogma of the radical separation of the Trinity's transcendent reality from the world of creation.

Yet, that does not make the church's existence apolitical but rather transforms the very idea of what constitutes public virtue and public action in the time that remains (between the Incarnation and the Parousia) (see also Williams 1987). What Peterson denied was not the possibility of the public representation of Christianity; that would have been entirely in opposition to his conception of the church as a visibly public (*öffentlich*) and social body. For Peterson, the public representation of the eschatological kingdom of God requires rejecting any and every attempt to identify this kingdom with a secular empire, principle of representation or liberal governmentality. Instead, the church's public presence is affirmed in its 'representative function testifying for the Incarnation and the promise that "God may be all in all" (1 Cor. 15:28)' (Gereby 2008, 24). The church as the body of the faithful lives and acts in history, but its guiding principle is founded on embodying the eschatological 'already but not yet' (*corpus mysticum*) that resists its appropriation by theological or secular myths of sacralised power.

Despite Peterson's often supersessionist (anti-Jewish) tendencies,[4] it is here that he comes the closest to Jewish theocratic messianism. As Gereby reminds us, 'the Old Testament concept of the Jewish kingdom explicitly resists the monotheistic temptation, too' (2008, 29). In 1 Samuel 9 the establishment of the kingdom for the Jews is requested based on analogy to the Gentile nations and *against* the will of God. God permits the Jews to elect a king, but this is a decision of the elect nation to which God grants his dispensation as an act of concession to a nation lacking faith (in God and in themselves);[5] it can serve as no confirmation of the sacrality of power or as a representation of the divine. No political – or, as it is more common in our times, technoscientific – utopia can masquerade as an earthly New Jerusalem outside of the final dystopia of the Antichrist. In both its Jewish messianic and Christian Trinitarian renditions, then, theocratic eschatology becomes not a progressive and revolutionary doctrine of destruction at the end of time but a discourse of critique and delegitimation of the powers of this world. The truth of eschatology – and, by extension, of hope as a theopolitical virtue – is disclosed in the umbilical cord that connects it to the soul of the Law (Decalogue), the critique of idolatry.

## Hope in the Posthumanist Apocalypse

At the early twenty-first century, it seems that theology, philosophy or political theory are increasingly seen as not having much purchase on 'ultimate matters'. What used to be entirely out of human hands appears practically possible, and threatens, or promises, to bring an end to the world as hitherto known. These possibilities include unprecedented devastation often evoking the apocalyptic imaginary: the annihilation of the world through nuclear weapons, depletion of natural resources, or human-induced climate change;

the destruction of cultures and forms of life through economic, technological or biomedical disasters spreading globally in a closely interconnected world. Conversely, the same possibilities promise a Promethean evolutionary leap: the attainment of an earthly eschatological utopia through death-erasing human enhancement or artificial super-intelligence (see Bostrom 2014). Alongside biotechnology, the progress of digital technology that made conceivable the development of artificial intelligence is heralded as 'the next stage in evolution' in which organic and artificial life enter a zone of indistinction and the limitations of organic existence are overcome in a state of posthumanist *eudaemonia* (Edward Fredkin quoted in McCorduck 2004, 401).

Meanwhile, at the precipice of such hitherto unimaginable possibilities, the traditional Christian message of salvation as eternal life has been dismissed by many as the premodern, mythical promise of a conceivable secular achievement. For secular eschatologists, genetic manipulation and other forms of biotechnological enhancement, if extended to human subjects, would enable them to transcend human boundaries, correct imperfections and cross the final frontier that used to be the privileged domain of consolation assigned to the religious sphere, mortality (see Manzocco 2019; Savulescu and Bostrom 2009). The unrelenting advance of such possibilities has engendered a widespread and inchoate sense of apocalyptic urgency: a collective feeling of having crossed the threshold to an unprecedent and unfamiliar terrain of history that is bound to redefine the ways human beings relate to politics, technology, art, common culture, life and death.

In this climate, twenty-first-century eschatological thinking seems to be shaped by visions of technoscientific millenarianism driven by the challenges of biology, cosmology and technology, rather than by the questions of theology, history and philosophy. If the familiar image of eschatology is that of an anti-political discourse, or a discourse too apolitical and too disconnected from the vagaries of history (emphasising the 'not yet'), the contemporary confluence of apocalyptic imaginary and technoscientific futurism pushes towards an 'applied' eschatology that eliminates the eschatological tension (prioritising the 'already').

Both Jewish messianic nihilism and Christian Trinitarianism aim to explode this dualism by showing how eschatology invites forms of genuine public engagement and critique of the powers that be, without being conventionally political. In modern takes of Jewish messianic and Christian Trinitarian theocracy, religious nihilism becomes a type of counter-politics that transforms the very idea of what politics stands for: neither the politics of sovereign or revolutionary violence nor the technoscientific effacement of politics, but rather the counter-politics of happiness, resistance, messianic profanation and theocratic an-*archy*. In this perspective, hope as a theopolitical virtue is affirmed within a terrain where politics and theology are no longer separate or juxtaposed discourses and where a certain nihilist take on the theological[6] is not only always already political, transforming the latter from within, but perhaps also the last stand (the true *katechon*) in a world wrapped up in apocalyptic frenzy.

## Notes

1. As such, Benjamin's outlook is anti-Machiavellian par excellence.
2. On the *nouvelle théologie* movement and the theology of the Second Vatican, see Flynn and Murray (2012).
3. Incidentally, Peterson, who was fully aware of these genealogies in patristic thought, became one of the inspirational reference points of *nouvelle théologiens*, such as Yves Congar, Henri de Lubac, Gaston Fessard and Jean Daniélou, who were part of the groundbreaking *ressourcement* movement in Second Vatican Catholic theology (see Shortall 2021).
4. See Peterson (2011, 104–5): 'only on the basis of Judaism and paganism can such a thing as "political theology" exist'.
5. See 1 Samuel 8:7–8: 'And the Lord told him: "Listen to all that the people are saying to you; it is not you they have rejected, but they have rejected me as their king. As they have done from the day I brought them up out of Egypt until this day, forsaking me and serving other gods, so they are doing to you.' As a *dispensatio*, the idea of kingship in Judaism and later the royal *regimen* in medieval Europe, at least until the rise of political Aristotelianism in the twelfth and thirteenth centuries, comes under the authority of divine *regnum*: 'But if you do not obey the Lord, and if you rebel against his commands, his hand will be against you, as it was against your ancestors' (1 Sam. 12:15). Rather than serving as some representation of the divine, monarchy is instituted as an act of disobedience that needs to remain under divine law to avoid regression to tyranny (see Senellart 1995).
6. For such a creative confluence of secular affirmative nihilism and Christian apophaticism, see also Newheiser (2019).

## References

Agamben, G. 2005. *The Time That Remains: A Commentary on the Letter to the Romans*. Stanford, CA: Stanford University Press.

Benjamin, W. 2007a. *Reflections: Essay, Aphorisms, Autobiographical Writings*. P. Nemetz (ed). New York: Schocken.

Benjamin, W. 2007b. Theses on the philosophy of history. In *Illuminations: Essays and Reflections*. H. Arendt (ed). New York: Schocken, 253–64.

Bielik-Robson, A. 2019. *Another Finitude: Messianic Vitalism and Philosophy.* London: Bloomsbury.
Bostrom, N. 2014. *Superintelligence: Paths, Dangers, Strategies.* Oxford: Oxford University Press.
Buber, M. 1990 [1932]. *Kingship of God.* London: Humanities Press International.
Flynn, G. and Murray, D. P. (eds). 2012. *Ressourcement: A Movement for Renewal in Twentieth Century Catholic Theology.* Oxford: Oxford University Press.
Gereby, G. 2008. Political theology versus theological politics: Erik Peterson and Carl Schmitt. *New German Critique* 105(3): 7–33.
Hollerich, M. 2011. Introduction. In E. Peterson, *Theological Tractates.* M. Hollerich (ed). Stanford, CA: Stanford University Press, xi–xxx.
Hotam, Y. 2007. Gnosis and modernity: A post-war German intellectual debate on secularisation, religion and 'overcoming' the past. *Totalitarian Movements and Political Religions* 8(3–4): 591–608.
Jonas, H. 1992 [1934]. *The Gnostic Religion: The Message of the Alien God and the Beginnings of Christianity.* London: Routledge.
Lazier, B. 2008. *God Interrupted: Heresy and European Imagination between the World Wars.* Princeton, NJ: Princeton University Press.
Lerner, A. 2015. *Redemptive Hope: From the Age of Enlightenment to the Age of Obama.* New York: Fordham University Press.
Lilla, M. 2007. *The Stillborn God: Religion, Politics and the Modern West.* New York: Vintage.
Löwy, M. 1992. *Redemption and Utopia: Jewish Libertarian Thought in Central Europe: A Study in Elective Affinity.* Stanford, CA: Stanford University Press.
Manzocco, M. 2019. *Transhumanism – Engineering the Human Condition: History, Philosophy and Current Status.* Cham: Springer.
McCorduck, P. 2004. *Machines Who Think,* 2nd edn. Natick, MA: A. K. Peters.
Newheiser, D. 2019. *Hope in a Secular Age: Deconstruction, Negative Theology, and the Future of Faith.* Cambridge: Cambridge University Press.
Paipais, V. 2018. 'Already/not yet': St Paul's eschatology and the modern critique of historicism. *Philosophy and Social Criticism* 44(9): 1015–38.
Paipais, V. 2022. Democratic political theology or divine democracy? *Political Theology* 23(3): 252–8.
Peterson, E. 2011. *Theological Tractates.* M. Hollerich (ed). Stanford, CA: Stanford University Press.
Phillips, E. 2015. Eschatology and apocalyptic. In C. Hovey and E. Phillips (eds), *The Cambridge Companion to Christian Political Theology.* Cambridge: Cambridge University Press, 274–95.
Savulescu, J. and Bostrom, N. (eds). 2009. *Human Enhancement.* Oxford: Oxford University Press.

Schmidt, C. 2007. Review of Jacob Taubes, *The Political Theology of Paul*. *Hebraic Political Studies* 2(2): 232–41.

Schmitt, C. 2003 [1950]. *The Nomos of the Earth in the International Law of the Jus Publicum Europaeum*. New York: Telos Press.

Schmitt, C. 2006 [1922]. *Political Theology: Four Chapters on the Concept of Sovereignty*. Chicago, IL: University of Chicago Press.

Scholem, G. 1971. *The Messianic Idea in Judaism and Other Essays on Jewish Spirituality*. New York: Schocken.

Schweitzer, A. 1968 [1893]. *The Quest of the Historical Jesus: A Critical Study of Its Progress from Reimarus to Wrede*. New York: Macmillan.

Senellart, M. 1995. *Les Arts de Gouverner: Du Regimen Médiéval au Concept de Gouvernement*. Paris: Seuil.

Shortall, S. 2021. *Soldiers of God in a Secular World: Catholic Theology and Twentieth-Century French Politics*. Cambridge, MA and London: Harvard University Press.

Styfhals, W. 2019. *No Spiritual Investment in the World: Gnosticism and Postwar German Philosophy*. Ithaca, NY and London: Cornell University Press.

Taubes, J. 2004. *The Political Theology of Paul*. Stanford, CA: Stanford University Press.

Taubes, J. 2009 [1947]. *Occidental Eschatology*. Stanford, CA: Stanford University Press.

Terpstra, M. and de Wit, T. 2000. 'No spiritual investment in the world as it is:' Jacob Taubes' negative political theology. In I. N. Bulhof and L. ten Kate (eds), *Flight of the Gods: Philosophical Perspectives on Negative Theology*. New York: Fordham University Press, 320–53.

Vatter, M. 2021. *Living Law: Jewish Political Theology from Hermann Cohen to Hannah Arendt*. Oxford: Oxford University Press.

Weiss, J. 1971 [1892]. *Jesus' Proclamation of the Kingdom of God*. R. H. Hiers and D. L. Holland (eds). Philadelphia, PA: Fortress.

Wight, M. 1948a. Christian commentary. BBC radio broadcast, 29 October.

Wight, M. 1948b. The Church, Russia and the West. *Ecumenical Review: A Quarterly* 1(1): 25–45.

Wight, M. 1955. Review of Eric Voegelin, *The New Science of Politics*. *International Affairs* 31(3): 336–7.

Williams, R. 1987. Politics and the soul: A reading of the City of God. *Milltown Studies* 19–20: 55–72.

Wolfe, J. 2022. Hope. In P. Ziegler (ed.), *The Edinburgh Critical History of Twentieth-Century Christian Theology*. Edinburgh: Edinburgh University Press, 333–44.

# Hope: An Epilogue
*Fleur Johns*

'The drama's done. Why then here does any one step forth? Because one did survive the wreck' (Melville 2015 [1851]). So ends Herman Melville's modernist classic, *Moby Dick*, with the story of a lone survivor 'drawn towards the closing vortex' but then 'upward burst' on a 'black bubble' that was 'floating on the margin of the [. . .] scene'. Is this how one should end a volume such as *Hope in the Anthropocene* – buoyed perhaps by the 'dark hope' of which Valerie Waldow, Pol Bargués and David Chandler write in this book's Introduction? It is difficult to know how to close a volume at once so promising and so insistent on holding readers open to irredeemable loss. This book invites readers to live without ever being 'picked [. . .] up' as Ishmael was at the end of Melville's tale.

Survival is certainly more apt a concluding note than summary for this fascinating book. No compendium could ever do justice to the many castings of hope in its pages. Never has a book been so crammed with hopes: hopes salvaged, repurposed, reoriented, dark, active, deep, abiding, problematic, false, crisis-riven, lost, educated, visualised, tempered, melancholic, unoptimistic, agentic, fanatical, foolish, courageous, conservative, governmental, compromised, refreshed, colonial and colonising, hegemonic, unevenly distributed, fugitive, draped in black, recomposed, speculative, pragmatic, nihilist, utopian, paradoxical, grounded or ungroundable, radical, strange, affective, unmoored, uncanny, dried-up, implicated, impossible, theopolitical, messianic, antinomian. Adjectival chaperones lead hope this way and that, bury and then elevate it. In hope's very diffusion, the questions that this book puts to readers become more specific than those with which evocations of hope typically travel: not *whether, when, why* or *what* to hope, but *how, under what conditions* and *compared to which alternatives*?

Of the alternatives surveyed in this book – hope's dancing partners, if you will – *critique* is a recurrent figure or theme, as the book's introduction rightly emphasises (and on which Chris Zebrowski's and Valerie Waldow's Chapters 2 and 11 also lay stress). At many points throughout the book hope seems to rise or to be salvaged from the ashes of that critical repertoire bequeathed to many (though not all) by Western Enlightenment thought. Yet hope is contrasted here to other dispositions, actions and affects too, among them belief, faith, humanism, wilfulness and optimism. To this book's contributors, hope is not reducible to any of these. Instead, hope seems to offer many of this book's contributors a way of marking and reorienting their work in the face of epochal shifts: from the linear times of classical physics and progressive international law to the simultaneity of quantum physics and Afrofuturism (in Geoff Gordon's Chapter 9); from concern for individual bodies to the experiencing of bodies as entangled and interactive (in Srishti Malaviya's Chapter 13); from the exhaustion of the 'postpositivist paradigm' and the 'fetishism' of entanglement towards articulations of a 'beyond-the-relational' (in Ignasi Torrent's Chapter 14); from 'rational' to 'complexity science' (in Claes Tängh Wrangel's Chapter 7); from the predominance of the 'major' to the centring of 'minor' gestures (in Chapter 3 by Sukanya Podder and Raúl Zepeda Gil and again in Srishti Malaviya's Chapter 13); from 'pandemic fatigue' to its managerial appropriation through 'presentist rhythmicity' (in Nicolas Gäckle's Chapter 6); from modernist ontology and a politics preoccupied with revelation towards a range of efforts to give the slip to the human/nature divide (in David Chandler's Chapter 1); and from both 'traditional theology and its secular humanist counterpart' to 'twenty-first-century eschatological thinking' (in Vassilios Paipais's Chapter 15). On this much the book's various contributors agree: proceeding according to business as usual as a scholar of social, political or legal theory or science is not an option, or at least not an option likely to have the purchase that work in these fields typically aims to have. For many, to write of hope as a central, animating concern – rather than as a morsel thrown begrudgingly to the reader at the end of a text – is to declare that most (if not all) of the classical repertoires of modern scholarship have expired or are facing imminent extinction in the Anthropocene.

What might it entail, then, to 'survive th[is] wreck' without trying to secure the future? The modes of survival canvassed in this book are noteworthy in part for what they are not. Survival is not mere persistence. As its etymology suggests, it involves something more than maintenance of life or affirmation of 'the world as it exists' (to quote from Chandler's Chapter 1). There is nothing debased, acquiescent or passive about it. It is active, collaborative and tactical. Unlike Ishmael in *Moby Dick*, surviving is not done alone. It is the people of Adjuntas, Puerto Rico, asking 'what next?' in a community meeting (in Christie Nicoson's Chapter 5 in this volume). It is the writing and reading of dystopian novels (in Aristidis V. Agoglossakis Foley's Chapter 12). It is W. E. B. Du Bois working with a group of students to conduct a study and craft a graphic depiction of the social lives of black communities in Georgia for the 1900 World Fair (in Kiran K. Phull's Chapter 4). It is the improbable staying power of what Geoff Gordon terms 'barred Black subject' (see Chapter 9). It is the insistence of Chief Plenty Coups (*Alaxchíia Ahú*) that the people of the Crow Nation 'shall somehow survive' in 'a new context' (per Jonathan Lear, quoted in Valerie Waldow's chapter in this volume). It is the Yanomami people's coexistence with nonhuman life forms in the forest (according to Renan Porto's chapter) even while Marjo Lindroth and Heidi Sinevaara-Niskanen in Chapter 8 caution that a 'hope–colonialism nexus' often serves to nullify Indigenous peoples' claims in the present.

While the temporalities of hope in which this book trades are multiple, it is also notable how much of this volume lingers in *the present* rather than vaulting towards the future. Hope is classically future-oriented, as many of the book's authors underscore. Nonetheless, most of the chapters are concerned with hope that is already here, already in the course of being practised or brought about. Admittedly, Gäckle's, Zebrowski's and Tängh Wrangel's chapters suggest that the presentism of contemporary hope may itself be problematic insofar as it fuels appetites for a return to normality, for managing and being managed, and for reconfiguring and securing the status quo. Yet the present of this book cannot easily bear these efforts of stabilisation, restoration or reunification given the multiplicity and simultaneity of presents that its chapters evoke.

In contrast to this book's overall orientation in and towards the present, epilogical scripts are, like gestures of hopefulness, often future-oriented. Consider George Orwell's epilogue to *Animal Farm*, for instance, Octavia Butler's ending to *Kindred*, or Charlotte Brontë's final aside to the reader in *Jane Eyre*. These endings entail speculation more 'predictive' than 'multiplicative'. That is, they 'narrow down multiple future possibilities to a single' future (Mancuso 2021, 461). No such concluding gesture is open to an epilogue of this volume, however, so I will not attempt it. The approaches to the future of this book's contributors are too variegated to bear winnowing to a single point. And the book's contributors themselves refrain from any such winnowing. Hopes' futures are not tomorrow-times in this book. They are the various ways that people are understanding, sensing and practising *now* what might stand beyond modernist reason and degradation.

In *Moby Dick* Ishmael is borne to life aboard Queequeg's coffin – or rather the oxymoronic 'coffin life-buoy' originally built for Queequeg on the premonition of his death and later repurposed after his recovery. Ishmael is rescued by a ship, the *Rachel*, looking not for him, an 'orphan', but for some cherished others: 'her missing children'. So too this book locates agentic hope in practices that are paradoxical, not masterful, purposeful or pre-scripted, and which proceed in and from destruction, even contempt. I write this epilogue on unceded Bidjigal land, in what is now called Australia, where the Munanjahli and South Sea Islander scholar Chelsea Watego has recently called for a 'retiring [of] hope for nihilism', not out of pessimism, but out of 'interes[t] in a Black life that is lived' and 'certain[ty] that there is a Black existence and a Black future available to us, beyond the present binary of white lies and Black deaths' (Watego 2021). Watego reads hope down before dispensing with it (as 'a waiting for a future good while living in a permanent hell'). The authors in this book give hope far greater amplitude than Watego does. Nonetheless, there is something of Watego's confoundingly joyous call to 'fuck hope' in the pages of this volume, in its strangely enlivening, negative invocations of 'hope against hope' (in the words of Chris Zebrowski's Chapter 8). The drama is ongoing. Survival is not assured. Yet the ending will not be in the spirit of Melville's, which bears a biblical epitaph suggestive of reconciliation and revelation. Many are making something else of loss and hope than disavowal, faith or delusion, as this book shows.

**References**

Mancuso, C. 2021. Multiplicative speculations: What we can learn from the rise and fall of hopepunk. *ASAP/Journal* 6(2): 459–83.

Melville, H. 2015 [1851]. *Moby Dick; or The Whale*. London: Penguin.

Watego, C. 2021. *Another Day in the Colony*. Brisbane: University of Queensland Press.

# Index

Aboriginal and Torres Strait Islander peoples, 144
adaptation, 46, 105, 113, 157, 161
Adorno, T., 4, 5, 6–7, 30–1, 154, 185
affect, 218, 219–20, 221
   loss of, 104, 113
   management, 45
   theory, 189, 218
affirmation, 6, 13, 32, 111, 114, 187, 191, 194, 196, 238, 253, 263
Afrofuturism, 263
Afrofuturist approaches, 26–7
Afropessimism, 30, 32
Agamben, G., 104–5, 251
agency, 1, 2, 3, 10, 27, 51, 52, 53, 77–8, 80, 130, 172, 175, 187, 190–1, 194, 237, 247, 252, 254
   German Development Agency, 61
   human, 1, 10, 189, 248, 250, 252
   Liberian Agency, 60
   youth, 54, 57
Agrilogistics, 7–8
Ahmed, S., 2, 93, 96
Anderson, B., 14, 106, 217, 218, 220–1, 222, 224, 226
*Annihilation*, 27
Anthropocene, 2–3, 7–9, 10, 11, 13–14, 21–2, 23, 24–5, 28, 33, 41, 42, 47, 104, 121, 150, 158, 168, 170–1, 186–7, 190, 191–2, 193, 194, 195, 196, 201, 208, 217, 229, 231, 232, 233, 235, 237, 238, 239, 241, 243–4, 247, 248, 263
   advent of the, 46–7
   different from the past, 42
   epoch of the, 33, 39, 40, 41
   geological epoch, 42, 171
   thinkers, 5, 10
anthropocentric, 25, 222, 229, 236
   non-, 33
*anthropos*, 171
*anthropy*, 171
'antiblack metaphysics', 31
anti-Black violence, 2, 107
apocalypse, 40–, 41, 247, 248, 256
apocalyptic, 40, 113, 235, 247, 248–9, 252, 255, 256, 257, 258, 259
Appadurai, A., 137
architecture, 159
artificial intelligence, 159, 235, 258

assemblage(s), 8, 29, 189, 190, 231
Auschwitz, 4, 31
*autogestión*, 87, 91, 92, 93, 94
Ayewa, C., 152, 153, 154

Baccolini, R., 202, 204, 205, 206
Barad, K., 161, 218, 224–5, 227, 228–9, 238
Barth, K., 249, 250
Battle-Baptiste, W., 82
becoming
  affective, 217
  complex and contingent, 105
  non-linear, 46
  politics of, 26
  process(es) of, 120, 225, 228, 235
  relations of, 10, 220
  Strange Becomings, 223–5
Benjamin, W., 29, 30, 250, 251, 253, 254, 255
Bennett, J., 25
Bering-Porter, D., 82–3
Berlant, L., 1, 44–5, 47, 48, 109, 151, 160
Bey, M., 154, 155, 162
beyond the human, 22, 119
Big Data, 29
biopolitical(ly), 105, 110, 112, 122, 129
biopolitics, 29, 105, 121
Bird Rose, D., 28, 29
Black
  being, 153, 154, 155, 156
  freedom, 143
  life, 68, 69, 71, 74, 78, 82, 153, 265
  people(s), 1, 27, 143, 167
  Quantum Futurism/BQF, 13, 27, 150, 152, 153, 154, 158, 160, 161, 162
  social theory, 150, 153
  studies, 30, 32

subject(s), 13, 24, 76, 82, 153–4, 155, 156
  *see also* bodies
  *see also* subjecthood
*Black Utopias*, 27
Bloch, E., 5, 6, 43, 192, 202, 203–4, 206, 213
bodies, 93, 107, 124, 130, 170, 171, 195, 218–20, 221–2, 223, 224, 226, 227, 228, 229, 263
  Black, 152, 156, 171, 174
  racialised, 82, 153
bottom-up, 11, 29, 126, 129
Bourdieu, P., 237
Bratton, B., 29, 30, 105
Brazil, 139, 167, 176
Breen, R., 52
Brown, J., 27
Buber, M., 250, 251–2
Burke, A., 238–9
Bussey, M. P., 55

*cabruca*, 176
Cahill, J., 52, 55
*Cannibal Metaphysics*, 29
capitalism, 3, 29, 46, 169, 170, 171, 174, 175, 189, 208, 239
  racial, 13, 24, 30, 33
*capitalocene*, 171
care, 33, 86, 87, 88, 89, 91, 93–4, 95, 96, 99, 138, 143
  as an epistemic tool, 87
  childcare, 56
  healthcare, 86, 88, 89, 95, 143
  networks of, 86
  solidarity and care, 113
  *see also* ethic of care
Cartesian dualism, 235, 236
cartography, 78, 178
Cassegård, C., 3, 8

catastrophe(s), 5, 6, 8, 12, 30, 70, 173, 175, 176, 195, 254
   climate, 12, 42
   colonial, 88
   ecological, 201, 208
causality, 23, 122, 190
   linear, 29, 119
Césaire, A., 124
Chandler, D., 11, 53, 120, 239
Christian, 39–40, 211, 248, 249, 250, 251, 252, 256, 257, 258
chronocenosis, 160–1
*chthulucene*, 172
Chua, D. K. L., 6–7
Cladis, M. S., 70
Clark, T., 8
climate, 96–7, 158, 167, 209, 258
   action, 86
   catastrophe, 12, 42
   change, 24, 37, 46–7, 55, 70, 86, 87, 88, 92, 167, 168, 170–1, 191–2, 208, 257
   crisis, 114, 212
   denial, 1, 24
   transformation, 92, 97
Colebrook, C., 31, 32, 113, 238
collapse, 28, 42, 152, 193
colonialism, 24, 30, 53, 86, 89, 141–3, 144, 145, 146, 169
   contemporary, 135, 142
   hope-colonialism nexus, 146, 264
   settler, 89, 142
   *see also* neocolonialism
   *see also* postcolonialism
Colonial Time of Hope, 144–6
coloniality, 3, 13, 97; *see also* decoloniality
colonisation, 86, 178
colour line, 69, 77–8, 80, 82–3
   global, 68, 69, 71, 74, 78, 83

complex
   life, 120, 125
   system(s), 39, 125–8
Complex Man, 120, 128, 130
complexity, 21, 25, 29, 40, 96, 119–21, 122, 124–5, 126, 127, 128, 129, 130, 226, 243, 245
   science(s), 119–20, 121, 124, 126, 127–8, 130, 263
   theory, 120, 125
composition(s), 173, 175, 176, 178, 190, 218, 219, 221, 224, 234
conflict(s), 51, 53, 54, 56, 62, 78, 160–1, 169
   ethnic, 138
   post-, 54, 57, 62
conformity, 207
Connolly, W. E., 25, 26, 238
conservation, 91, 161
consumerism, 207
correlationism, 25
cosmology, 233, 236, 258
   relational, 236
cosmotechnics, 178
counter-mapping, 79–80
courage, 9, 202, 208, 209–10, 211, 212
COVID-19, 37–8, 45, 47, 103, 104, 105, 106
   pandemic, 38, 103, 105, 106, 107, 108, 109, 113–14
   polycrisis, 39, 47
   response, 45, 47
crisis(es), 37, 38, 39–41, 42, 43, 44–5, 46, 70, 104, 107, 108, 140, 145, 187, 190, 192, 196, 223, 248, 250
   climate, 114, 212
   economic, 39, 51
   of hope, 38, 192
   ordinariness, 45, 47, 109
   suicide, 138, 143
   *see also* polycrisis

critical theory, 5, 10, 186, 189, 191
critique(s), 31, 37, 38, 40, 42–3, 45, 46, 47, 48, 87, 90, 130, 143, 151, 154, 155, 172, 185–7, 188–91, 192, 195, 196, 211, 220, 237–8, 239, 247, 248, 250, 257, 258, 263
   of modernity, 119, 121, 153, 194
   posthumanist, 119, 121
*Critique of Cynical Reason*, 43
Crow Nation, 9, 139, 264
*Cruel Optimism*, 44, 151
Crutzen, P., 171
*Curandeiro Trabalhando Com Tabaco*, 178
cybernetic(s), 120–1, 168
cynicism, 38, 43–4, 46, 47, 48, 188

Danowski, D., 172, 170
data, 68, 69, 71, 72, 74, 75, 76, 78, 80, 81, 82–3, 158, 159, 170
   futures, 69, 80, 83
   subjects, 82
   visualisation, 68, 72, 83
   *see also* Big Data
Davidson, J. P. L., 70
Davison-Vecchione, D., 202
decolonial
   epistemology and politics, 121
   *scienta*, 119, 123, 130
decoloniality, 121; *see also* coloniality
decolonizing the mind, 27, 28
deconstruction, 3, 33, 236
Deleuze, G., 26, 220, 225, 232, 234
democracy, 37, 135
denial, 1, 24, 55, 153, 155
depoliticising, 237–8, 239
despair, 30, 43, 54, 80, 111, 139, 187, 204, 206, 207, 208, 209, 213, 221, 228
destruction, 5, 13, 24, 28, 32, 139, 140, 169, 173, 204, 205, 208, 248, 257, 258, 265

   environmental, 33, 39, 171, 178
   total, 14, 206
detachment, 81, 110, 114, 250, 252
development(s), 39, 42, 53, 54, 55, 59, 70, 92, 95, 136, 137, 139, 141, 142, 158, 159, 176, 180, 185, 194, 235, 248, 249, 258
   business, 59
   economic, 88
   global, 41, 137, 139
   intervention, 57, 62
   policy (policies), 53, 57
   post-war, 51
   urban, 89
*Dialectic of Enlightenment*, 4, 5
dichotomy, 77, 204
difference(s), 7, 24, 27, 31, 42, 56, 72, 78, 81, 82, 106, 107, 159, 161, 162, 171, 206, 221, 222, 225, 226, 234, 252
Dinerstein, A. C. 139
disaster, 22, 28, 46, 193, 258
disavowal(s), 3, 13, 24, 33, 237, 265
discourse(s), 1, 2, 11, 12, 13, 24, 25, 30, 40, 107, 120, 130, 137, 150, 160, 168, 173, 192, 218, 225, 235, 236, 243, 247, 250, 257, 258, 259
   contemporary, 170, 180
   medical, 39, 106
   military, 121
   peacebuilding, 58
disenchantment, 3, 174
dispossession, 2, 24, 144, 153
disruption(s), 3, 33, 37, 106, 159, 161, 162, 171, 220, 221, 224
divides between humans and nonhumans, 109
domination, 27, 32, 80, 88, 145, 175, 196, 202, 207, 209, 212
double consciousness, 31

Du Bois, W. E. B., 11, 31–2, 68, 69, 70–1, 72, 74–6, 77, 78, 80, 81, 82–3, 264
Du Boisian spiral, 74
dystopian
  criticism, 213
  studies, 202, 204
  thought, 201, 202, 204, 206, 212

Eagleton, T., 145
Earth, 7–8, 10, 23, 32, 170, 173, 175, 176, 180, 235
  degradation, 175
  surface, 171
  systems, 23
ecocide, 13
ecologisation, 168
ecosystem(s), 93, 95, 158, 161, 175, 180, 191, 231
Edelstein, D., 160–1
emergency, 12, 45, 95, 103, 104–8, 110–11, 113
  governance, 103–4
  present, 103–4
  response, 45, 104, 106–7, 111
emergency/everyday distinction, 107
encounter(s), 9, 26, 78, 103, 219, 220, 229, 233–4, 239, 240, 242
end
  of history, 249, 188
  of the world, 21, 26, 32, 40, 172, 173, 194, 238, 247–8
  of time, 247, 255, 257
endurance, 104, 109, 111, 112, 114
energy, 12, 61, 91, 92, 95–6, 171
  solar, 91, 94, 95
  systems, 86, 89, 92, 94
Enlightenment, 4, 5, 11, 21, 22, 23, 40, 43, 46, 48, 121, 130, 188, 263

entanglement(s), 39, 109, 119, 141, 160–1, 162, 191, 194, 217, 218, 223, 227–8, 231–2, 233, 236, 237–8, 239–40, 241, 243, 244, 263
  fetishism, 232, 237, 240, 243
  fetishists, 232, 239–40
  ontological, 240, 242
  pandemic, 112
entropy, 167, 168, 169, 170, 174, 175
  cosmic, 170, 173, 175
  environmental, 168, 175
  psychic, 170
  social, 168, 170
environment, 10, 21, 47, 59, 87, 88, 92, 96, 126, 127, 129, 140, 158, 168, 173, 176, 218, 222, 247
episteme, 81, 87, 93, 96
Ernstson, H., 237
Esbell, J., 178
eschatology, 14, 247–8, 249, 250, 256, 257, 258
ethic of care, 87, 92–3, 94, 95, 96
ethics, 10, 42, 69, 191, 192
ethnocentrism, 172
Eurocentric, 25, 29, 30, 234
Eurocentrism, 186
Europe, 40, 110, 159, 176, 233
European Commission's Quantum Flagship initiative (EQF), 3, 150, 158, 160, 161–2
EU's Digital Decade, 154, 157, 158
experimentation, 11, 25, 82
exploitation, 21, 70, 80, 92, 138, 211
extinction, 8, 9, 10, 28, 41, 180, 208, 263
extraction, 10, 136, 138, 175
extractivism, 2, 11, 88

Fanon, F., 153, 154
fascism, 3, 54
fatigue, 103–4, 108, 109–11, 112, 113–14
  collective, 103
  managing lockdown, 103
  pandemic, 103, 104, 107, 109, 110–11, 112, 113, 114, 263
feedback loops, 28, 29
Ferdinand, M., 171
Ferreira da Silva, D., 152, 154, 156
Forster, E. M., 205
fossil fuel(s), 86, 91–2, 96
Frankfurt School, 4, 25, 30
Freud, S., 5
Fromm, E., 25
future(s), 2, 5, 6, 10, 13, 14, 21, 26–7, 31, 37, 42–43, 51, 52, 53–4, 55–6, 59, 62–3, 69, 70, 71, 74, 75–6, 80, 82, 83, 86, 87, 90, 91, 92, 93, 94–5, 97, 103–4, 108, 110, 113, 130, 135, 137, 145, 146, 152, 158, 160, 162, 167–8, 170, 172–3, 180, 193–4, 195, 201, 206, 222, 243, 264–5
  hopeful, 54, 63, 141, 160, 243, 244
  positive, 56, 204
  youth, 53, 54–6, 62–3
futurity, 1, 72, 112, 145, 154, 155

Gaia, 175
Garland, A., 27
genetic manipulation, 258
genocide, 13, 88
geoengineering, 172, 180
geological forces, 8
geology, 7
Gesturing Towards Decolonial Futures, 10
Gilroy, P., 32

Glissant, È., 241, 242
globalisation, 23, 137, 174
global warming, 8, 12, 167, 235
Gnosticism, 250, 252, 254
Goldthorpe, J. H., 52
governance, 22, 24, 29, 38, 103, 105, 109, 111, 113, 120, 121, 124, 125, 127, 129, 130, 150, 157, 158, 160, 161, 189, 190, 194
  crisis, 107
  emergency, 103, 104
  global, 120, 154, 239
  imaginaries of, 110, 113
  liberal, 104
  pandemic, 103, 104, 108, 112
  planetary, 29
  security, 120, 126
  self-, 87, 91, 97
Grosz, E., 121
Grove, J., 3, 25–6
Guattari, F., 167, 175, 225
Gumbs, A. P., 9

Hage, G., 137, 138, 139–40
Haraway, D., 168–9, 172, 231
Harney, S., 153, 154, 155, 157, 160, 161, 162
Haro, L., 137, 139
Hartman, S., 81
Hehir, A., 2
Hesiod, 4
Hine, D., 3
horizon, 24, 80, 104, 105, 108–9, 110, 111, 113–14, 220–1, 232, 248
  beyond the, 223, 226
  normative, 189–90
horizonlessness, 103–4, 109–10
Holocaust, 3, 4, 5, 6, 7, 9, 10, 185; *see also* Auschwitz

hope, 1, 2, 3–4, 5–6, 7, 9, 10–12, 13–14,
    21–3, 24–5, 33, 37–8, 42–3, 42–5,
    47–8, 51, 52–3, 56, 69– 71, 72,
    74, 76, 80, 81, 83, 86–7, 91, 92,
    103–4, 106–7, 108, 109, 110,
    111–12, 113–14, 119, 120, 121,
    122–3, 124, 130, 135, 136–8,
    139–41, 142, 143, 144–6, 150,
    154–5, 157, 160, 161, 162, 167–8,
    170, 176, 180, 185, 186–7, 188,
    191–2, 193–5, 196, 202–3, 204–6,
    206–9, 210–12, 213, 217–18,
    220–1, 222, 223, 224–5, 226–7,
    228–9, 231–2, 243, 244, 247–8,
    254, 257, 262–3, 264, 265
  affective, 14, 222, 226
  against hope, 38, 47–8, 208,
    265
  agentic, 71, 72, 74, 83, 265
  Aristotelian, 210, 211
  as a theopolitical virtue, 14, 248,
    257, 259
  as governance technology, 157
  as practice, 70
  Blochian, 203
  conservative, 107, 108
  courageous, 210, 211, 212
  cruel, 45
  cynical, 157, 158, 160
  dangerous, 251
  dark, 3, 7, 9, 14, 262
  decolonial, 121
  dialectic of, 69, 70, 80, 83
  draped in black, 150, 154, 155, 158,
    160, 161, 262
  educated, 87, 89, 90, 97
  encoded, 69
  eschatological, 248, 251
  false, 37, 207, 208–09, 254
  fanatical, 202, 203, 204, 205, 213

fresh, 51
fugitive, 150, 154, 155
global, 140
good, 2
ideological nature of, 24
Kantian, 27
loss of, 140
melancholic, 155, 161, 162
messianic, 251
modern, 24–5, 28
modernist, 23, 24, 47
more-than-human, 236
negative, 202, 203, 205, 206–9
nihilist, 11, 30–2, 33, 53–4, 70
non-being of, 6
ontology of, 26
paradox of, 3, 72
positive, 202, 203, 210
posthuman, 119
post-war, 52
pragmatic, 28–30, 31, 33, 53–4, 62,
    63
progressive, 108
progressive technological, 161
radical, 9, 139, 150, 187, 192, 193–4,
    196
redemptive, 253
religious, 247, 248, 250
revival of, 13, 186–8, 191–2, 193
secularised, 40
sowing seeds of, 91
speculative, 25–7, 28, 33, 53–4, 60,
    62, 63
statistical, 69, 71
tenacious, 155
tentative, 83
true, 207, 254
unconscious-as-, 5
unhopeful, 69–70
unmoored, 218

unmooring of, 222, 229
utopian, 192–3, 195, 196, 204
Western, 141
hopefulness, 136, 137, 140, 142, 145, 146, 213, 264
hopeless, 11, 47, 69, 74, 80, 109–10, 126, 137, 208, 209, 253
hopelessness, 38, 44, 54, 55, 103, 108, 109–10, 139, 142, 207, 254; *see also* resignation
*Hospicing Modernity*, 3
Hui, Y., 175, 178
human(s), 6, 7–8, 9, 22–3, 27, 28–9, 31, 32, 40, 61, 70, 76, 78, 88, 92, 95, 109, 119, 120, 121–2, 123, 128–9, 130, 156, 168, 170–2, 175, 179–80, 186, 189–90, 191–2, 193, 202, 203, 204–5, 208, 209, 219, 222, 229, 234, 235–6, 242, 243, 251, 258
 exceptionalism, 7, 25, 238
 history, 8, 55, 186
 life, 6, 69, 81, 119, 123, 170–1
 mastery, 5, 47
 new genre of the, 2
 rights, 135, 136
 society, 180, 208
human/nature
 binaries, 12, 25, 125
 divide, 25, 33, 263
human–nature relations, 187, 154
humanism, 122, 238, 248, 263
 new, 2
humanity, 4–5, 23, 42, 43, 135, 140, 146, 171, 172–3, 189, 205, 207–8, 209, 238, 239, 241–2, 247
Hutchinson, F. P., 55
*hutukara*, 173
Huxley, A., 207
Hyperobjects, 23

ideology, 3, 6, 18–19, 43, 173
imaginary (imaginaries), 2, 12, 26, 29, 32, 39, 52, 53, 54, 68, 71, 72, 75, 82–3, 89, 105, 107–8, 113, 127, 150, 158, 162, 169, 172, 188, 195, 217, 235, 241
 apocalyptic, 247–8, 257–8
 governmental, 10, 24, 103, 108, 109–10
 liberal, 104, 108, 110
 modernist, 9, 10, 23, 26
 of progress, 23
 political, 10, 87, 186, 188, 191
Inayatullah, S., 55
indeterminacy, 14, 194, 218, 223, 229
indigeneity, 139, 147, 148
Indigenous, 12, 29, 136, 138–46, 173, 176, 203
 communities, 29, 56, 138, 139, 140, 143, 144, 176
 cosmologies, 13, 168, 180
 dispossession, 24
 lands, 178
 peoples, 8, 12, 28, 29–30, 88, 135, 136, 137, 138–41, 142, 143–4, 145, 146, 171, 172–3, 264
 state relations, 142, 143, 144, 145, 146
 studies, 142, 143
 worlds, 172, 178
innovation, 9, 126, 159
 governmental, 105
 technological, 89, 201, 207
insecurity, 38, 39, 42, 47, 126, 127, 186, 194, 248
Intergovernmental Panel on Climate Change, 46
international law, 150–1, 153, 154, 239, 263

Johnson, J. M., 45
Johnston-Goodstar, K., 55–6
Joseph, J., 237
justice, 40, 46, 56, 68, 69, 83, 136, 143, 144, 195, 250, 251, 256, 262
 environmental, 12, 87, 89
 social, 89

Kandinsky, W., 82
Kant, I., 22, 23, 28, 185
Kern, A. B., 38, 42
Kingsnorth, P., 3
Klein, M., 15n
knowledge, 4, 8, 22, 23, 28, 42, 47, 72, 76, 82, 87, 90, 93, 105, 119, 121–2, 139, 169, 188, 189, 211, 224, 225, 238, 256
 ecological, 176
 Indigenous, 140, 173
 new order of, 119, 123, 130
 situated, 93
 traditional, 140
 unsituated, 119
 Western, 121
Kohn, E., 29
Kopenawa, D., 173, 178, 179–80
Koselleck, R., 38, 39, 40, 42, 46
Koskimaki, L., 56
Kurki, M, 231, 236

Lacan, J., 15n, 163n
Latour, B., 21, 29, 168, 175, 176, 188–91, 237
Lear, J., 8, 139, 140, 193, 264
learning, 9, 53, 112, 194
Leibniz, G., 232, 234
Levitas, R., 90
Lewis, J. S., 24, 123
liberal, 11, 21, 24, 29–30, 52, 107, 108, 110, 123, 124, 128, 169, 174, 190, 238–9, 242, 249, 257

liberalism, 24, 28, 54, 143, 151; *see also* neoliberalism
liminality, 6, 223
local, 11, 55, 57, 58, 61, 69, 71–2, 74, 76, 78, 83, 88, 89, 90, 91, 95, 96, 121, 126, 128, 211
loss (ontological), 2–3, 4

Machado de Oliveira, V., 2–3, 7
Macuxi people, 178
Malevich, K., 82
Manning, E., 218–19, 225–6, 227
Marcel, G., 43
marginalisation, 54, 142
Marxism, 154
Marxist, 90, 204, 211, 237, 239
 neo-Marxists, 297, 239
mass, 2, 9, 28, 88, 110, 157, 201
Massol González, A., 87, 91, 95–6
material, 29, 32, 178
 conditions, 52, 69, 196
 dimension, 243
 inequalities, 129
Mbembe, A., 174
McMullin, J. R., 62
Melians, 206, 208, 209
Melville, H., 262, 265
memory, 47, 125, 128, 152, 153, 208
 traumatic, 141
Mercator projection, 78
Middle Passage, 27, 155
 epistemology, 26
Mignolo, W., 122, 130
militarisation, 88, 121
modern
 civil society, 155, 156
 faith, 42
 life, 11, 223
 ontological cuts, 238, 239
 politics, 254

modernisation, 167–8, 170, 173–5, 176, 180
modernist, 2–3, 5, 6, 10, 11, 12, 13, 14, 22–3, 24–5, 29, 31, 33, 40–1, 129, 186, 187, 188, 189, 190–1, 195, 218, 264
  political thinking, 7
modernity, 2–3, 4, 6, 8, 11, 22–23, 24–5, 26, 30, 32, 33, 40–1, 70, 71, 72, 82, 119, 121–3, 130, 153, 156, 189–90, 192, 194, 201, 236, 247
  end of, 4, 162
  failures of, 233
  liberal, 30, 190
  violence and unsustainability of, 7
  Western, 171
modes
  of becoming, 231
  of being, 12, 27, 175, 232
  of consciousness, 123
  of entangled being, 28
  of governance, 103, 109, 125, 150
  of life, 226
  of production, 171, 174
  of subjectivation, 112
  of survival, 263
Mondrian, P., 82
Moore, J., 171
more-than-human(s), 13, 29, 88, 92, 95, 180, 236
Morin, E., 38, 40, 41–2
Morton, T., 8, 23
Moten, F., 153, 154, 155, 156, 157, 160, 161, 162
mourning, 9, 13, 155, 156, 158, 160
Muñoz, J. E., 87, 90
Murphy, B. J., 76–7, 80, 81

narrative(s), 8, 11, 24, 40, 45, 46, 48, 63, 69, 70, 71, 72, 74, 76, 78, 80, 81, 87, 168, 170, 190, 192, 193, 201–2, 205, 207, 208, 210, 212–13, 221, 222, 235, 243, 252
  progress, 1, 26–7, 151, 158
natural, 5, 21, 22, 77, 93, 121, 140, 201, 223, 236
  events, 170, 171
  resources, 92, 257
nature, 4–5, 7–8, 9, 21, 22, 23, 28–29, 30, 70, 121–2, 123, 124, 125, 129, 140, 142, 143, 172, 175, 186, 190, 205, 208, 210, 223, 236, 243
  domination of, 5
nature/culture divide, 8, 237
necropolitics, 169
negative, 2, 6, 13, 72, 76, 127, 186, 196, 203, 206–7, 210, 247, 250, 255, 265
*Negative Dialectics,* 7
*negrocene,* 171
neocolonialism, 211
neoliberalism, 28, 37, 109, 174
  governmental rationality of, 38
new materialism, 189
Newtonian, 22, 125
nihilism, 14, 30, 31, 62, 242, 248, 251–4, 258, 265
  messianic, 14, 254, 258
nonhuman life, 119, 123, 180, 264
non-linearity, 113
normality
  new, 38
  return to, 12, 38, 45, 47, 48, 106, 107, 110, 113, 264
*nouvelle théologie,* 256

Ogawa, Y., 208
ontology, 11, 26–8, 31, 33, 112, 156, 186, 190, 199, 231–2, 234, 237, 239, 242, 243, 244, 263
   flat, 237
   modern, 21–2, 25, 28, 29
   modernist, 7, 29, 31, 263
   of hope, 26
   relational-materialist, 237
   viral, 105
*operaista*, 169
oppression, 26, 70, 76, 78, 152, 202, 205, 207, 210, 212, 239, 242, 244
opportunity, 37, 46, 52, 53, 54, 120, 137, 142, 190, 194
optimism, 44, 55, 136, 151, 192, 194, 249, 263
   cruel, 38, 44, 47, 138, 150, 151, 155, 157, 158, 160, 161, 162, 192
   quantum, 154, 158
Overbeck, F., 249

Pandora's box, 4
paraontological
   disruption, 33
   unravelling, 33
*Partial Connections*, 30
*perichoresis*, 256
Peterson, E., 256–7
Phillips, R., 152, 160, 161
philosophy, 4, 90, 191, 231, 233–4, 235, 257, 258
*Physics of Blackness*, 27
planet politics, 238–9
planetary, 8, 170, 171, 174, 180, 194, 231, 239
plantation(s), 88, 171
   logics, 1

plurality, 53, 227
pluriverse, 175, 243
policy (policies), 3, 11, 12, 23, 24, 53, 57, 61, 62, 63, 89, 106, 157, 158, 160, 161, 162, 238, 243
politics of hope, 150, 158
polycrisis, 38–9, 40, 41–2, 47
   epoch of the Anthropocene, 39, 40–1
Pospisil, J., 51
postapocalyptic
   politics, 3
postcolonial, 24, 56, 83, 121, 136, 141–2, 143, 144, 146, 170, 211, 213
postcolonialism, 141
posthuman, 119, 120, 121, 190
posthumanist, 119, 121, 257–9
Povinelli, E., 24, 29–30, 109, 143
power(s), 12, 24, 26, 28, 29, 37, 45, 46, 70, 71, 72, 79, 82, 83, 104, 114, 127, 135, 136, 137, 138, 141, 150, 153, 160, 162, 169, 185, 187, 192, 196, 203, 206, 208, 210, 248, 250–1, 255, 256, 257
   anarchical, 252
   colonial, 145, 241
   economic, 89
   immanent, 22
   liberal, 29
   Messianic, 254
   political, 97, 254
   relations, 78, 120, 124, 138, 236
   sacralised, 257
   secular, 249
   sovereign, 247
   structural, 76

presentism, 264
problematisation, 3, 23, 42, 104, 110, 111, 113, 153, 224
production, 2, 8, 9, 12, 55, 61, 72, 94, 96, 142, 151, 157, 169, 175, 176, 226, 238
progress, 2, 3, 4, 5, 7, 9, 23–24, 32, 40, 42, 43, 69, 70, 71, 74, 76, 78, 80, 111, 135, 145, 146, 155, 157, 161, 162, 167, 172–3, 187, 194, 225, 235, 247, 254, 258

quantum, 12, 27, 154, 158–60
  entanglements, 218, 227–8
  phenomena, 158
  physics, 152, 160, 263
  theory, 152
  see also Black Quantum Futurism
  see also European Commission's Quantum Flagship initiative

racism, 39, 47, 70, 171
racist, 70, 144, 156
  anti-, 78, 83
radical openness, 14, 231, 232, 243, 244
radicalisation, 68, 127, 128, 129, 249
Rational Man, 119, 120, 121–2, 123, 128, 130
rationalisation, 174
rationalist, 26, 189, 238
realism, 53, 238
  political, 206
recomposition
  ecological, 13, 167, 168, 175
reconciliation, 7, 31, 58, 136, 141, 143, 144, 241, 265
reflexivity, 90
regulation, 23, 128, 195

relationality, 161, 189, 190, 231, 232, 233, 235, 236, 237, 238, 240, 242, 243
religion(s), 185, 234, 247, 252, 260
repression, 122, 156
resignation, 1, 56, 70, 250, 255
resilience, 3, 15, 45, 46, 54, 109, 115, 118, 164, 197
resistance, 9, 14, 30, 70, 72, 88, 91, 96, 110, 124, 139, 154, 156, 205, 210, 211, 213, 244, 258
  political, 14, 88, 232, 241, 242
*resonances*, 192
revelation, 4, 40, 45, 248, 263, 265
rhythm(s), 108, 152, 170, 225, 227
risk, 11, 14, 23–4, 46, 54, 60, 62, 81, 120, 137, 157, 180, 237, 242
Roitman, J., 41
ruins, 3, 8, 31, 249

salvation, 3, 6, 7, 9, 11, 13, 40, 42, 48, 205, 206, 208, 247, 249, 253, 258
Santos, M., 173–4
Sargent, L. T., 204
*Savage Ecology*, 26
Schmitt, C., 247, 252, 255
Scholem, G., 251, 252–3, 254
sciences, 105
  complexity, 12, 119, 120, 121, 124, 127, 128, 130
  earth system, 231, 235
  environmental, 158
  natural, 122
  neurosciences, 120, 126, 128
  psy-sciences, 103
  quantum, 158, 160
  social, 2, 74, 152, 188, 231
Scott, J. C., 22

securitisation, 136, 159
security, 8, 46, 47, 95, 111, 128, 136, 160, 174
  communication, 158
  cybersecurity, 159
  environment, 126
  governance, 120, 126
  insecurity, 38, 39, 42, 47, 126, 127, 170, 186, 194, 248
  quantum, 159
  water, 91
*Seeing Like a State*, 22
seer, 26, 27, 33
separation, 7, 76, 228, 236, 237, 248, 253, 256
slavery, 70, 76–7, 156, 171, 242
  abolition of, 68
  chattel, 24, 27, 33, 169
  institution of, 88, 89
Sloterdijk, P., 43
Smith, A., 22, 23
social mobility, 52, 53, 54, 55, 56, 59, 61, 174
sovereignty, 105, 153, 159, 161, 225, 251
  Anarchic, 252
  divine, 255
  technological, 159, 160
spacetime(s), 26, 173, 218, 219, 221–2, 224, 225, 226, 227–8, 234
species, 9, 24, 28, 39, 41, 115, 122, 172, 173, 176, 180, 190, 208, 231
speculation(s), 235, 264
Spinoza, B., 234
Stanley, S., 43
state(s), 43, 53, 69, 71, 72, 76, 105, 107, 138, 139–40, 141, 142–3, 144, 145, 146, 169, 170, 174, 188, 194, 195, 207, 210, 211
  absent, 95
  -centred, 139
  colonial, 139
  European, 161
  measures, 113
  mechanisms, 169
  neocolonial, 211
  neoliberal, 139
  of emergency, 105
  of exception, 104, 105, 252
  of posthumanist *eudaemonia*, 258
  of the world, 250
  regulation, 23
  sovereign, 160
  structures, 139, 143
  welfare, 170
  *see also* Indigenous-state relations
*Staying with the Trouble*, 30
Stengers, I., 236
Stewart, F., 55
strangeness, 14, 217–18, 222–4, 226, 228, 229
Strategic Research Agenda, 158, 159
Strathern, M., 30
subjecthood, 80, 81–2
  Black, 71, 72, 76, 78, 80
subjectivity, 108, 109–10, 111, 112–14, 153, 155, 170, 218, 220, 223
  Black, 153
  human, 14, 222, 229
  political, 189, 232, 241, 242
subject/object divide, 25
surveillance, 209
survival, 44, 52, 54, 86, 139, 169, 193, 232, 241, 242, 244, 262, 263, 265
Sustainable Development Goals, 57

sustainability, 52
*sympoiesis*, 172
*synoikēsis*, 256
system(s), 13, 28, 37–8, 46, 77, 86,
    94–5, 122, 125–9, 143, 152,
    156, 158, 171, 172, 176, 207,
    233
  legal, 141, 143, 153
  political, 250
  social, 155
  technical, 174–5
Swyngedouw, E., 237

Taubes, J., 254–6
technology (technologies), 55, 61,
    119, 158–60, 161, 168, 170, 172,
    174, 175, 205, 207, 210, 213, 220,
    258
  biopolitical, 112
  biotechnology, 258
  governance, 157
  neoliberal, 113
  of security, 119
  of security and war, 119
  quantum, 150, 158, 159
  security, 106
  studies, 235
technoscientific, 14, 172, 175, 235,
    248, 257, 258
teleological, 40, 41, 74, 241
temporality (temporalities), 10, 70,
    104, 105, 109, 139, 145, 161, 174,
    264
  linear, 24, 26, 162
  modernist, 41
  nonhuman, 114
  stretched, 107, 110
*The Comet*, 32
*The Ends of the World*, 172
*The Falling Sky*, 173

*The Long Emancipation*, 24
*The Principle of Hope*, 5, 192, 202,
    203
*The Revenge of the Real: Politics for a
    Post-pandemic World*, 29
*The Souls of Black Folk*, 69
*The Three Ecologies*, 167
*The Undercommons*, 157, 161
Thörn, H., 3
threat(s), 39, 44, 91, 96, 126, 127, 151,
    192
Thucydides, 202, 203, 205–7
time
  epiphenomenal, 26, 27
  geological, 171
  historical, 41, 108
  kairotic, 106–7
  linear, 26, 109, 151–3, 163
  postapocalyptic, 8
  trickery, 152
  viral, 103, 105, 109
Tooze, A., 39, 41, 42
totalitarianism, 201, 204, 210, 211,
    247
tragedy, 81, 154, 170, 172
Trinitarianism
  Christian, 14, 254–7, 258
Tsing, A. L., 30, 31
Tupinambá, G., 176, 178–9

uncertainty, 2, 23, 44, 54, 56, 105, 120,
    126, 137, 187, 228
  radical, 105
*Undrowned: Black Feminist Lessons from
    Marine Animals*, 9
United Nations Development
    Programme (UNDP), 57–8,
    60-1
United States, 55–6, 68, 72, 86, 88, 89,
    92; *see also* US

unpredictability, 23, 46, 126, 129, 187, 189
(un)settlements, 51, 53
urban, 59, 74, 76, 89, 140
urbanisation, 173
*urihi*, 173, 178, 180
US, 86, 88, 89, 120, 125
  military, 88, 119–20, 125, 126, 127, 128, 129, 130
utopia(s), 87, 90, 94, 96, 155–6, 257
  critical, 204
  concrete, 87, 89, 90, 92, 93, 96, 97
  eschatological, 258
  minor, 54–5, 56, 61–2, 63
  orientations of, 94
utopian, 86, 87, 88, 186, 187, 191, 192, 195, 196, 202, 203, 204
  dreaming, 8
  impulse, 202, 204, 205
  project(s), 48, 54, 90
  tradition, 204
  *see also* hope, utopian
utopianism, 90, 192

veil, 25, 31, 32, 69, 70
Vieira, P., 204
violence, 2, 7, 24, 30, 31, 39, 53, 54, 57, 69, 74, 80, 81–2, 107, 108, 109, 128, 129, 139, 141–2, 143, 152–4, 155, 156, 161, 170, 186, 239, 241, 242, 243, 247, 250, 258
  colonial, 80, 86, 239, 242
  logics of, 242, 243–4
  military, 170
  monopoly of, 169, 174
  racial, 69, 70, 78, 81, 82
  social, 90, 167
visualisation(s), 11, 54, 68–9, 71, 72, 76, 78, 80, 81, 82–3

Viveiros de Castro, E., 29, 168, 172, 176
vulnerability (vulnerabilities), 47, 86, 127, 138
  ontological, 187, 193, 194, 243

Walcott, R., 1, 2, 24, 143
Walsh, L., 56
Warren, C., 31
war(s), 51–2, 53, 54, 57–8, 119, 120–1, 122, 169, 250
  civil, 57, 62, 138
  Cold War, 188
  culture, 38
  First World War, 249–50
  Second World War, 208, 247
  Spanish American War, 88
ways of
  being, 95, 97, 146
  life, 140, 171, 172, 175
  living, 48, 94, 103, 174
Weheliye, A. G., 81, 82
*We Have Never Been Modern*, 29
West, C., 161
Western, 55, 68, 80, 121, 141, 151, 171, 174, 193, 196, 203, 208, 211, 213, 216, 238, 248, 263
  liberal societies, 52
  societies, 56
  tradition, 172, 233
white, 76, 78, 151, 156, 235, 242, 265
  Europeans, 172
  non-white, 167–8, 171, 175
  peoples, 174
  supremacism/supremacist, 70, 78, 82
woman, 32
world domination, 80

Whitehead, A N., 235
wicked problems, 8
Wight, M., 247, 248
Wilderson III, F. B., 153, 241–2
Williams, R., 203
Winters, J. R., 69, 154–5
Wolfe, J., 248
World Bank, 59
World Health Organization (WHO), 103, 104, 107, 109, 110, 111–12
World Wide Web, 55

Wright, M., 26–7, 151, 153, 162
Wynter, S., 119–20, 121–4, 126, 128, 130

Yanomami, 173, 178, 179, 180, 264

Zalloua, Z., 153, 155, 156
Zamyatin, Y., 201, 205, 210
Zapatista, 139
Zeus, 4
Zuberi, T., 81

EU Authorised Representative:
Easy Access System Europe Mustamäe tee 50, 10621 Tallinn, Estonia
gpsr.requests@easproject.com

Printed and bound by CPI Group (UK) Ltd, Croydon, CR0 4YY
02/03/2026
02063692-0012